1992

The Living Earth
The Coevolution
of the Planet and Life

Other Books in the
DISCOVERING EARTH
SCIENCE SERIES

VOLCANOES AND EARTHQUAKES (No. 2842)

The first book in the series, this volume concentrates on the geologic phenomenon of Earth, and how this phenomenon has affected life on our planet. Also included is a history of the planet's geology.

VIOLENT STORMS (No. 2942)

This second book in the series deals with the atmospheric and climatic phenomena of Earth, as well as their effect on man and their influence on our planet.

MYSTERIOUS OCEANS (No. 3042)

The third volume in the series looks at the hydrologic phenomenon of Earth—its origin, purpose, mechanisms, and effect on life. The book concentrates on oceans' role on our planet.

EXPLORING EARTH FROM SPACE (No. 3242)

The final volume in the series deals with the technologic advances that enable us to view our planet from space. It covers the way man-made satellites can provide us with better information on the geologic, climatic, biologic, and hydrologic phenomena of our planet and enable us to better predict disasters, locate and monitor natural resources, and explore our Solar System.

The Living Earth
The Coevolution
of the Planet and Life

JON ERICKSON

Discovering Earth Science

TAB BOOKS Inc.
Blue Ridge Summit, PA

FIRST EDITION
FIRST PRINTING

Copyright © 1989 by TAB BOOKS Inc.
Printed in the United States of America

Library of Congress Cataloging-in-Publication Data

Erickson, Jon, 1948-
      The living earth : the coevolution of the planet and life / by Jon
Erickson.
          p.   cm.
      Bibliography: p.
      Includes index.
      ISBN 0-8306-8942-7   ISBN 0-8306-3142-9 (pbk.)
      1. Coevolution. 2. Life—Origin. 3. Earth—Origin.
4. Biosphere. I. Title.
QH372.E75   1988
577—dc19   88-26557
                                                          CIP

TAB BOOKS Inc. offers software for sale. For information and a catalog, please contact TAB Software Department, Blue Ridge Summit, PA 17294-0850.

Questions regarding the content of this book should be addressed to:

   Reader Inquiry Branch
   TAB BOOKS Inc.
   Blue Ridge Summit, PA 17294-0214

Cover photograph courtesy of United States Geological Survey.

Edited by Suzanne L. Cheatle
Series design by Jaclyn Saunders

Contents

143,456

Introduction

IMAGINE going back 4.6 billion years and witnessing the birth of a living planet. The Earth formed out of dust and gases in the protoplanetary disk that circled the infant Sun. By the time the Earth had grown to nearly its full size, the Sun flared up and blew away its original atmosphere, exposing the raw surface, which was in constant turmoil with large meteor impacts and giant volcanic eruptions. Then out of the heavens came a Mars-sized asteroid that was kicked out of the asteroid belt between Mars and Jupiter by a passing comet. The asteroid struck the young planet with a mighty blow, and as it sped off into space, it drew a streamer of molten rock out of the planet's interior. Some of this material escaped into space and some fell back to Earth, but most began to orbit the planet and eventually coalesced into the Moon.

The unique properties of the Earth-Moon system caused tides in the oceans, and the tides as well as the Earth's temperature, which was between the freezing and boiling points of water, helped the Earth acquire life much earlier than had previously been thought possible. A great deal of carbon existed on the Earth, and organic compounds were produced by the reaction of atmospheric gases with powerful lightning bolts and intense ultraviolet light from the Sun. The oceans became a rich broth of complex organic molecules that combined by chemical reactions into amino acids, the very building blocks of life.

The first living entities were simple organisms that obeyed the most fundamental definition of life. Living cells became more complex as time went on, and some were able to obtain energy directly from the Sun. The conversion of solar energy into carbon compounds liberated oxygen into the ocean and atmosphere. The introduction of oxygen heralded the development of multicellular organisms, and environmental diversity produced a wide variety of species.

After the planet was some 4 billion years old, life crawled out of the ocean and onto the land. It might seem strange that it should have taken so long since life had been in existence for at least 3 billion years, but terrestrial organisms could not venture onto the land until the ozone layer developed in upper atmosphere. The ozone layer blocks out deadly ultraviolet radiation from the Sun.

Simple plants took the first tentative steps and were closely followed by insects, some of which evolved side by side with plants. The first amphibians were fish that were adapted to live off the land as well as the sea. The amphibians advanced into reptiles, which in turn evolved into dinosaurs. The mammals developed alongside the dinosaurs, and for still unexplained reasons, they survived the great Cretaceous extinction 65 million years ago, while the dinosaurs and 70 percent of all other species vanished. The mammals quickly became masters of the planet, and one particular species could walk upright, communicate, and build elaborate tools.

When this highly inquisitive animal advanced into space and could view the Earth from afar, he was forced to the conclusion that the whole of humanity lived on a fragile spaceship. No longer could humans wantonly ruin the land, destroy the forests, pollute the air and water, and cause the extinction of millions of species of plants and animals without having dramatic and irreversible effects on the Earth that would ultimately affect man and his offspring.

We are already seeing many repercussions of these activities. The oceans from which life originally sprang are losing much of their variability because of pollution and overfishing. Rain forests, which contain the greatest numbers of different species, are being felled at alarming rates to feed a hungry world. Life generally can adjust to most changes over long periods (measured on a geologic time scale), but sudden, extensive changes are usually followed by the extinction of whole species. The changes wrought by man fit into the latter category, and unless these destructive activities cease, the Earth might witness the greatest extinction event in its entire history, requiring millions of years to recover from the irresponsibility of a single species.

Acknowledgments

The following organizations are appreciated for their assistance in providing photographs for this book: the U.S. Department of Agriculture, the National Aeronautics and Space Administration, the National Oceanic and Atmospheric Administration, the National Park Service, Rushmore Photo Inc., the U.S. Coast Guard, the U.S. Geological Survey, the U.S. Maritime Administration, the U.S. Navy, and the U.S. Soil Conservation Service.

1

Making a Special Planet

THE story of life on Earth must start at the beginning of the universe. There are three basic theories dealing with the creation of the universe (FIG. 1-1). The Big Bang theory asserts that the universe started out dense and is becoming less dense as it expands. According to the Steady-State theory, the universe is expanding but new matter is being generated to maintain a constant density. The Pulsation theory (also known as the Big Bounce theory) states that the universe expands only so far, and then gravity pulls it back together into a dense state, where upon it reexplodes, creating a new universe. This book deals with the Big Bang theory.

According to the Big Bang theory, approximately 15 billion years ago (give or take a couple billion years), the universe was in a highly condensed state. There were no atoms, protons, or electrons, but only subatomic particles packed together in a dense cosmic soup with a temperature that was magnitudes greater than the interior of the largest star. As the protouniverse expanded, it cooled sufficiently to allow the basic units of matter to clump together and form the seeds of the galaxies. The simplest

atoms—hydrogen and helium—formed first and gathered into gas clouds. The gases condensed into globules by mutual gravitational attraction. Further compression caused the first stars to ignite.

Astronomers are able to glimpse back toward the beginning of the universe by observing what appears to be a massive protogalaxy in its formative stages. It is 12 billion light-years from Earth, meaning that the object is being seen as it existed only a few billion years after the Big Bang. About this time, the Milky Way Galaxy had pulled enough matter together to form a large spiral galaxy, similar to those observed in the far reaches of space (FIG. 1-2). The Milky Way is composed of 100 billion stars spread out across a distance of 100,000 light-years. With this many stars and lots of time to work with, it is conceivable that life, or at least its precursors, existed in our Galaxy long before the Earth did.

THE MILKY WAY

In the Northern Hemisphere on a clear, moonless night away from city lights, a band of stars can

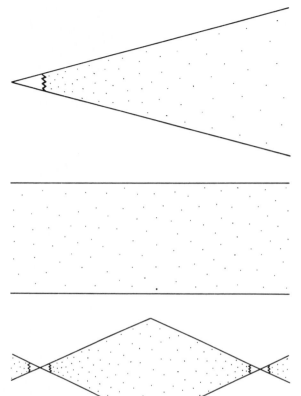

FIG. 1-1. Creation of the universe. (Top) the Big Bang, (middle) the Steady State, (bottom) the Big Bounce theories.

be seen stretching from northeast to southwest in the summer and northwest to southeast in winter. Early star gazers thought this broad ribbon of light was some sort of ethereal substance in the heavens. The early Greeks called it *galaxias kuklos*, which means "milky ring," hence came the name Milky Way Galaxy. It was not until the invention of the telescope by Galileo in the early 1600s that the Milky Way was resolved into an uncountable number of stars (FIG. 1-3.)

With the construction of large telescopes in the twentieth century, the structure of the Milky Way was finally revealed. Great aggregations of stars called *globular clusters*, which can pack up to 1 mil-

lion stars in a very small space, were found to have a high density in the constellation of Sagittarius. Radio astronomy pinpointed this region as the source of strong radio emissions, and powerful gamma rays and x-rays also radiated from this area. Therefore, Sagittarius is thought to be the center of our Galaxy.

The Milky Way has a central bulge with a radius of about 15,000 light-years and consists of densely packed old stars. It has been suggested that in the center of our Galaxy as well as the other galaxies, there exists a black hole up to 1 billion times the Sun's mass. It insatiably gobbles up stars, and not even light can escape its strong gravitational pull, which is why it is black. Outside the central bulge is the galactic disk wherein our solar system resides. It has a radius of 50,000 light-years and is composed of relatively young stars, as well as gas and dust. A *galactic halo* with a radius of 65,000 light-years has widely spaced old stars and roughly half the globular clusters in the galaxy. The combined mass of the bulge, disk, and halo is equal to the mass of about 300 billion suns. Beyond the galactic halo is the outermost component of the Galaxy, called the *corona*. It stretches out to about 300,000 light-years from the galactic center and contains old, burned out stars and poorly luminous objects called *galactic companions*, which include globular clusters, dwarf spheroidal galaxies, and the irregular galaxies that make up the Large and Small Clouds of Magellan.

The Galaxy has three principal spiral arms that peel off from the central bulge like the swirls of a huge vortex. These arms are regions of star formation. An interstellar medium of gas and dust within the spiral arms dims the light from the stars in the central plane of the Galaxy. The interior of these dark interstellar clouds appears to be where most of the Galaxy's new stars are born. In addition to hydrogen and dust, the clouds are also made up of a mixture of complex organic molecules.

The central plane of the Galaxy completes one revolution about the center roughly every 200 million years. Thus, our solar system, which is about 30,000 light-years from the center, is racing through space at a speed of 0.5 million miles per hour (the distance to the Moon and back). Present-day space-

FIG. 1-2. The Whirlpool Galaxy.

craft take several days to complete a trip of this distance.

Astronomers believe that the Milky Way and the Andromeda Galaxy, which is 2 million light-years away, together make up a *binary system* and are bound together by gravity and orbiting a mutual center of gravity. According to theory, our Galaxy is following an orbit in the shape of a narrow, flat ellipse that takes it away from Andromeda and will bring it back for a close encounter with its sister galaxy in about 4 billion years. If a collision between the two galaxies occurs, it could trigger a *quasar*, an exceedingly energetic neutron star, in the center of the Galaxy which could outshine the rest of

(Courtesy of NASA)

FIG. 1-3. The Milky Way Galaxy showing satellite trail.

the Galaxy. The intense radiation could boil away the interstellar gas clouds within our Galaxy, ending its ability to make new stars. The collision also could tear the Galaxy apart and shoot our Sun off into intergalactic space. Because the planets are so tightly bound to the Sun, they would simply hang on for the ride, and if somehow man survived, he would see an entirely different sky at night.

A STAR NAMED SOL

The ancient Greeks believed that the Sun occupied a special place in the cosmos, and even astronomers of the early twentieth century held the belief that the Solar System existed in the center of our Galaxy. It has only been within fairly recent years that scientists have come to grips with stellar evolution and the development of our Solar System. Two or three times a century, a giant star explodes somewhere in our Galaxy and becomes a supernova. Because of their great size, these giants burn them-

selves out within only a few million years, and the last stages become an explosive event that rips the star apart.

Cassiopeia A is a remnant of a supernova that was first observed in the year 1680 and lies some 9000 light-years from Earth. The fireball is presently about ten light-years across, and some dense blobs of gases that dot the fireball are as massive as 300 Earths. The expanding shell of hot gases is punctured by stellar fragments, giving the supernova a mottled appearance. These fragments might be the reason supernovas stay so bright for so long.

More recently, on February 23, 1987, the first known supernova near enough to be visible from the Earth in four centuries was observed in the Large Magellanic Cloud, less than 200,000 light-years away. It was named simply 1987A.

Once every few years, a new star is born in our Galaxy. These stars originate from an assortment of nebulas, molecular complexes, and globules, all of which are composed of condensing clouds of gas and dust (FIG. 1-4). A typical globule has a radius of about one light-year and a mass of about 100 Suns. In addition to hydrogen and helium, it also contains organic compounds such as carbon monoxide, formaldehyde, and ammonia. The time it takes for a complete collapse to form a star is only about 1 million years. The density wave from a nearby supernova provides the outside pressure needed to start the collapse (FIG. 1-5). The density wave partially envelops the globule and compresses it, causing a rapid collapse by self-gravitation. As the wave passes through, spiral arms peel off, forming a protoplanetary disk that looks much like a miniature spiral galaxy.

The image of the disk that caused the Sun to be born can be still traced from the motions of the planets in our solar system. All the planets orbit the Sun in the same direction with near circular orbits; all except Venus rotate in the same direction; and all except Pluto move in the same plane, called the *ecliptic*.

In our Solar System, debris from the supernova mingled with the original gas and dust, and these materials were incorporated in the construction of the Solar System. Late arrivals condensed independently into a primitive class of meteors composed of carbon, called *carbonaceous chondrites*, which are totally alien to our Solar System. The center of the protoplanetary disk collapsed faster than the periphery, causing a nodule to form. The collapse converted the infalling gravitational energy into kinetic energy, and the temperature of matter in the center rose high enough for thermonuclear reactions to begin, signaling the birth of the star. When a fairly large star was created its radiation blew away any remaining gas and dust that surrounded it, and it became a lonely giant. Multiple star systems of two or more stars, which make up about 80 percent of the stars in our Galaxy, formed when the nodule fissioned. The stars orbit each other around a common center of gravity and are not believed to have planets.

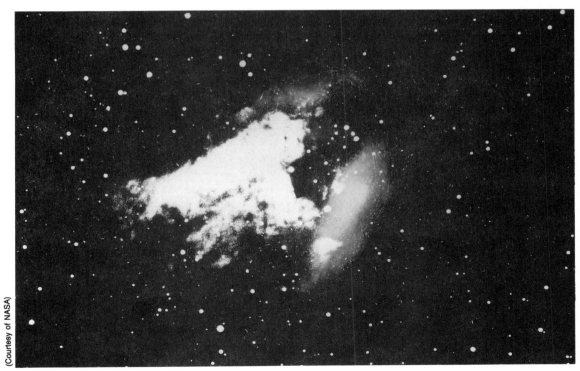

(Courtesy of NASA)

FIG. 1-4. The Swan Nebula.

The Sun is an average-size star about one-third as old as the Milky Way Galaxy. During the first billion years of its life, the Sun was highly unstable, periodically puffing itself up by as much as one-third larger and casting off gigantic solar flares that reached millions of miles into space. The *solar wind*, composed of subatomic particles and responsible for the tails of comets, was more like a solar hurricane compared to what it is today.

The Sun has a layered structure, consisting of

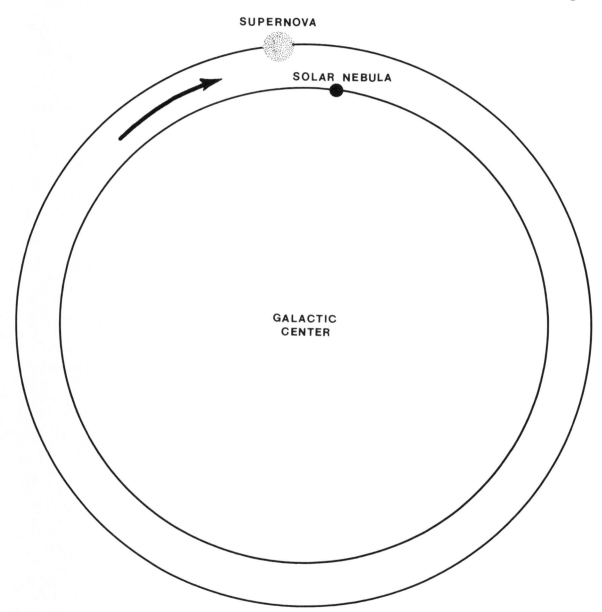

FIG. 1-5. Star formation from the collapse of the solar nebula.

an inner *core* of helium, an intermediate layer of hydrogen called the *radiative zone*, and an outer layer of hot gases called the *convection zone*. The *photosphere* is the visible surface of the Sun and has a temperature of 6000 degrees centigrade. It is often marred with *sunspots* which are large patches of relatively cool gas associated with magnetic vortices and energetic solar flairs. The *chromosphere* is the Sun's atmosphere and is composed of reddish glowing gases thousands of miles thick. An intensely hot halo called the *corona* extends millions of miles into space.

The Solar System is quite large and consists of nine known planets. Scientists speculate there is a small unseen planet inside the orbit of Mercury, to which they have given the name *Volcan*, for the Roman god of fire. Another undiscovered large planet, called *Planet X*, is thought to have an eccentric orbit, inclined to the ecliptic beyond the orbit of Pluto, and appears to have had some gravitational influence on Uranus and Neptune. Because Pluto's orbit is inclined 17 degrees to the ecliptic, it is thought to be a captured planet or possibly a moon of Uranus knocked out of orbit by a comet.

Between the orbits of Mars and Jupiter lies the asteroid belt, and one asteroid was so large it was once mistaken for another planet. The asteroids are leftovers from the formation of the Solar System that were unable to coalesce into a planet, probably because of the gravitational tug of war between Jupiter and the Sun. They come in a variety of shapes and sizes up to 600 miles wide.

Seven billion miles from the Sun is the *heliopause*, which marks the boundary between the Sun's domain and interstellar space. About 20 billion miles from the Sun is a region of gas and dust, possibly remnants of the original solar nebula.

Surrounding the Solar System about a light-year away is a collection of millions of comets known as the Oort cloud, named for the Dutch astronomer Jan Oort who first postulated its existence. Many comets have highly elliptical orbits and they swoop out of the heavens, swing close to the Sun at fantastic speeds, and fly back out into space again. Some of them are not so lucky and plunge straight into the Sun, where they are vaporized, apparently having

no adverse effects on the Sun. The Sun's nearest neighbor, Alpha Centauri, is about 4 light-years away; and at the speed of present-day spacecraft, it would take 100,000 years to reach it.

THE PROTOPLANETS

Orbiting the Sun during its early stages of development was a protoplanetary disk composed of several bands of coarse particles called *planetesimals*. From the primordial dust, the particles grew through weak electrical and gravitational attractions, as they swung around the Sun in elliptical orbits, which resulted in constant collisions. The planetesimals closest to the Sun were composed of stony and metallic minerals, and ranged from the size of sand grains to huge bodies over 50 miles across, but most were about the size of a pebble. Farther from the Sun, where temperatures were much colder, solid chunks of water ice, frozen carbon dioxide, and crystalline methane and ammonia formed. If it were not for the presence of large amounts of gases in the Solar System, which tended to slow down the planetesimals and caused them to spiral in toward the Sun, the larger bodies might have swept up all the smaller planetesimals, producing a solar system composed of rings like those of Saturn instead of planets.

It appears that protoplanetary disks that might eventually coalesce into planets are fairly common accoutrements of young stars. The disks of dust can be detected indirectly because they cut off some of the out-flowing stellar wind and disrupt its velocity, like rocks disrupt the smooth flow of water in a stream.

In the early stages, the inner terrestrial planets (Mercury through Mars) had similar developmental histories. A metallic core composed of ferromagnetic elements such as iron and nickel might have formed first by mutual magnetic attraction of metallic planetesimals. After the core was fully developed, gravitational forces were strong enough to attract the stony planetesimals, which piled up layer by layer on top of the core. The planet also could have formed undifferentiated with a homogeneous mixture of rock and metal. High-intensity radioactive elements,

Table 1-1. Classification of the Earth's Crust.

ENVIRONMENT	CRUST TYPE	TECTONIC CHARACTER	CRUSTAL THICKNESS	GEOLOGIC FEATURES
Continental crust overlying stable mantle	Shield	Very stable	22 miles	Little or no sediment; exposed Precambrian rocks
	Midcontinent	Stable	24 miles	
	Basin & Range	Very unstable	20 miles	Recent normal faulting, volcanism, and instrusion; high mean elevation
Continental crust overlying unstable mantle	Alpine	Very unstable	34 miles	Rapid recent uplift; relatively recent intrusion; high mean elevation
	Island arc	Very unstable	20 miles	High volcanism; intense folding and faulting
Oceanic crust overlying stable mantle	Ocean basin	Very stable	7 miles	Very thin sediments overlying basalts; no thick Paleozoic sediments
Oceanic crust overlying unstable mantle	Ocean ridge	Unstable	6 miles	Active basaltic volcanism; little or no sediment

called *radionuclides*, generated a tremendous amount of heat, which melted the planet from the inside out. The planet then separated into concentric layers, with the heavier metallic elements falling toward the center and the lighter substances floating toward the surface. The infalling planetesimals also heated the surface by impact friction. The entire process took only about 100 million years, and resulted in a planet that was partially melted throughout.

The presence of a gaseous medium that pervaded the early Solar System gave each of the terrestrial planets a massive atmosphere, and its gravitational compression created surface temperatures greater than the melting point of rocks. The formative Sun at first glowed red hot as gases compressed the core. Then suddenly, the Sun flared up as the core ignited in a thermonuclear reaction. This produced a strong solar wind that eroded away the original atmospheres of the terrestrial planets and transported the gases out to the giant gaseous planets (Jupiter through Neptune) and beyond.

The outer planets have rocky cores as large as the Earth if not larger, a mantle possibly composed of ice, and a thick layer of compressed gas—mostly hydrogen, helium, methane, and ammonia. Pluto, the moons of the outer planets, and the comets are essentially rock encased in thick layers of ice. Jupiter has about the same composition as the Sun, and if it had not stopped growing, it could have ignited like a small companion star and rivaled the Earth's moon in brightness.

After the Sun cleared away the gases in the vicinity of the terrestrial planets, they became orbiting spheres of hot barren rock, devoid of any atmosphere, like Mercury and the Moon are today. Without an atmosphere, the surface of each planet rapidly cooled, forming a primitive crust like the film on a bowl of cold pudding. This was not a true crust, for the interior of each planet was still in a molten state, and agitation by convection currents kept the mantle well mixed and did not allow chemical separation. Therefore, the density of the rocks solidifying on the surface was not much different than that of the mantle. As a result, the crust was very unstable, and it either remelted on the surface, dove back into the mantle and remelted, or simply overturned like a top-heavy boat and remelted. This instability is the reason there is no geologic record of the first 700 million years of the Earth's existence.

The smaller terrestrial planets, including the Moon, cooled more rapidly than Earth, allowing separation between the mantle and crust to take place much sooner. As convection currents became more sluggish, which resulted from the loss of radiogenic heat as radionuclides mutated into stable daughter products, lighter rock materials were allowed to migrate to the surface and form a permanent crust. Between 4.2 and 3.9 billion years ago, a massive meteor shower peppered the planets and their moons. Mercury, Venus, Mars, and Earth's Moon, as well as the moons of the outer planets, show numerous pockmarks from this invasion and little has happened since. Earth had not yet developed a permanent crust, so the meteorites simply splashed into a thin layer of scum and kicked up quantities of partially solidified or molten rock. The scars quickly healed over with fresh lava, and no telltale signs of the craters can be found today.

THE BIG SPLASH

The origin of the Moon still remains one of the biggest puzzles in science. Moon rocks brought back by the Apollo astronauts in the early 1970s are believed to be similar in composition to the Earth's upper mantle. The rocks range in age from 3.2 to 4.5 billion years. Since no rocks were found dating younger than 3.2 billion years, the Moon probably ceased volcanic activity at that time, and its interior began to cool and solidify. The oldest rocks, called *Genesis Rock*, formed the original lunar crust and originated deep within the Moon's interior.

Several theories have been put forward to account for some of the baffling qualities of the Moon. The Moon was once only about half the distance it presently is from the Earth; therefore, it can be argued that the Moon was plucked out of the Earth either by centrifugal force when Earth was spinning rapidly on its axis or by tidal forces generated by a passing asteroid. The fact that the Moon's composition is somewhat different from that of Earth has led to the belief that it was a passing asteroid captured by the Earth. Its large size, only slightly smaller than Mercury, has provoked speculation that the Moon and Earth together are a twin planetary system, both evolving together while orbiting each other around a common center of gravity.

A new model of lunar evolution better explains how the Moon came into existence and combines the best ideas of the older theories into a single cohesive hypothesis. While the Earth was still in its molten state, it was constantly bombarded by comets and asteroids, some as wide as 50 miles. The large impacts sent rocks flying 100 miles or more into space, but because they could not achieve escape velocity, the debris simply fell back to Earth. Therefore, it required a much larger impact in order to place enough material into orbit around the Earth to form the Moon. Unfortunately, such a collision might have been detrimental to the Earth, possibly shattering it and scattering its remains far and wide.

An alternative theory envisions an asteroid about the size of Mars striking Earth but with a glancing blow (FIG. 1-6) instead of a direct hit thereby gouging out a substantial amount of material without destroying Earth in the process. As the asteroid rebounded into space, it pulled along a streamer of vaporized rocks from Earth's interior, as well as some of its own rocks, which also vaporized from the giant explosion. Some of this material was lost in space, and some returned to Earth, but a substantial amount went into orbit around Earth and formed a prelunar disk.

Just as Earth and the rest of the planets formed out of the protoplanetary disk, the Moon grew by the accretion of material in the prelunar disk until all the debris in its eccentric path around the Earth was swept clean. In the process, the Moon heated up by radioactivity, compression, and impact friction so that it became a molten orb revolving around the Earth. With a mass of about one-eightieth and a volume of about one-fiftieth that of Earth, the Moon quickly cooled and formed a permanent crust long before the Earth did. Thus, during the great meteor bombardment that ended 3.9 billion years

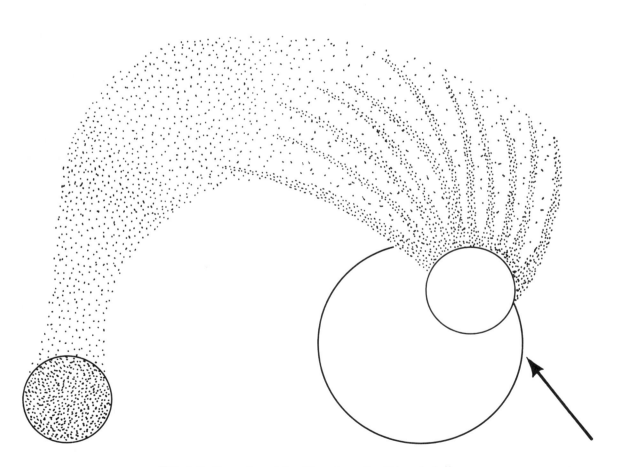

FIG. 1-6. Formation of the Moon from the "big splash."

ago, the Moon became highly cratered and developed much of the terrain features it has today (FIG. 1-7).

The Moon orbited Earth so closely that it filled much of the sky and caused huge tidal bulges in the Earth's crust, which was just forming. The Earth's rotation rate was much faster too, and days were just a few hours long. As the Earth transferred some of its angular momentum to the Moon, its rotation rate slowed down, while the Moon moved progressively farther from the Earth. It is possible that, by having a nearby moon, or *sister planet* as it is some-

FIG. 1-7. The Earth's Moon.

times called, to produce tides in the oceans, Earth acquired one of the most important properties needed for the creation of life.

THE INNER EARTH

As Earth began to cool and convection currents in the interior began to slow down, the planet was able to complete its differentiation into core, mantle, and crust. The *core*, a little over half the diameter of Earth, makes up about one-sixth of its volume and about one-third of its mass. The inner core, composed of iron-nickel silicates, solidified into a sphere about 1500 miles in diameter. The outer core, composed mostly of iron, remains liquid at a temperature ranging from 4500 degrees Celsius at the top to 7000 degrees Celsius at the base, and flows as easily as water.

It is because of this structure that the core is able to generate a strong magnetic field. Convective motions in the fluid core set up a dynamo effect, whereby electrical currents generate magnetic fields, which are reinforced by the slowly rotating solid core. Periodically, on the order of about a half million years or so, and for unexplained reasons, the geomagnetic field collapses and is regenerated in the opposite direction. The surface of the fluid core is not smooth, but has a topography composed of mountains as tall as Mount Everest and valleys as deep as the Grand Canyon which are formed by the rising and sinking of the mantle by convection currents. The sloshing of the jagged core against the mantle also could make the Earth's rotation jerky, first speeding up and then slowing down by tiny fractions of a second.

The Earth's mantle comprises nearly half the radius, 83 percent of the volume, and 67 percent of the mass of the planet. Mantle rocks are composed of iron-magnesium silicates in a particularly molten or plastic state, which makes them flow like thick molasses. This ability to flow enables large-scale convection currents to transport heat away from the core and in turn gives rise to small-scale convections, which distribute the heat along the surface of the mantle. The mantle is divided into a lower layer from the top of the core to about 400 miles

beneath the surface and is composed of primitive rock unchanged since the early development of the Earth. The rocks of the upper mantle have been actively changing in composition and crystal structure since the beginning. It is for this reason that no exact matches of moon rocks with mantle rocks can be made, since the Moon has been tectonically dead for over 2 billion years.

Plumes of molten rock from deep within the lower mantle and called *hot spots* rise to the surface and create volcanic islands and volcanoes in the interior of continents. In this respect, all the activity on the Earth's surface is just an outward expression of the great heat engine that drives the mantle. Continents separate and collide, mountains rise and trenches sink, and volcanoes erupt and faults quake. The mantle played an important role in shaping the Earth and giving it a unique character. Large amounts of gases and water vapor sweated out of the mantle through volcanic eruptions in what geologists term the "big burp." The mantle provided the atmosphere, the oceans, the land, and the carbon from which life was created.

THE UPPER CRUST

About four billion years ago, the Earth began to develop a permanent crust. The oldest rocks found on Earth are from the 3.8 billion-year-old Isua formation in southwest Greenland (FIG. 1-8) and are composed of water-lain, metamorphosed sediments, indicating the surface was cool enough for oceans to form. When the mantle began to cool, the stirring of hot rocks in this vast cauldron became sluggish, allowing the separation of rocks into lighter and heavier components. The lighter rocks, or *scum*, floated to the surface like a slag heap. The crust is therefore the main repository of elements incompatible with the mantle; in other words, refuse the mantle did not want. Oxygen, silica, and aluminum make up the bulk of the crust, forming the granitic rocks of the continents.

The upper brittle crust and the upper brittle mantle, together called the *lithosphere*, are separated by a layer of ductile rock, giving the crust a

FIG. 1-8. Location of the Isua formation in southwest Greenland.

thick buoyant crust would have remelted because of the high concentration of radioactive elements and the great pressures induced by the weight of the overlying rocks. A thick crust would act as an insulating blanket and not allow the constantly generated heat from the Earth's interior to escape into space, therefore raising the internal temperature high enough to melt the crust. Cooling also would have made a thick crust highly unstable, resulting in a massive overturn that would have melted the crust. The fact that there is no vestige of the first 700 million years might attest to these occurrences early in Earth's history. A thick, buoyant crust could not be easily broken up and subducted into the mantle, which is important for global plate tectonics, and plates would simply float on the surface like pack ice in the Arctic Ocean. This would indeed make the Earth an uninteresting place. There would be no majestic mountains, no deep valleys, no volcanoes and earthquakes, and no life.

The first land to appear were the *cratons*, which are found in the hearts of all continents (FIG. 1-9). They are composed of ancient igneous and metamorphic rocks and are remarkable in that they are very similar in composition to recent rocks. The existence of these rocks also might be evidence that plate tectonics operated early in Earth's history. Sediments thrust deep into the mantle are subjected to the Earth's high internal heat. The rocks either change their crystal structure or melt entirely and become magma. The buoyant magma rises toward the surface in blobs called *diapirs*. If the magma breaks through the surface, it produces volcanic eruptions. Otherwise, it remains buried in the crust to form granitic rocks.

The cratons were free-wheeling slivers of crust that collided and bounced off each other. As the Earth continued to cool, the cratons slowed their erratic wanderings and began to stick to one another, forming over a dozen protocontinents ranging in size from smaller than the state of Texas to about one-fifth the area of present-day North America, and together they comprised only one-tenth of the total landmass of Earth today.

Continental collisions crumpled the crust, forcing up mountain ranges at the point of contact (FIG.

structure like a jelly sandwich. The lighter rocks of the crust ride like icebergs on the fluid rocks of the *asthenosphere*, which is the molten outer layer of the mantle. Only the tips of the continents above the oceanic crust, which is composed of the basaltic rocks of the lithosphere.

The Earth's crust is relatively thin compared with the Moon and the other terrestrial planets. A

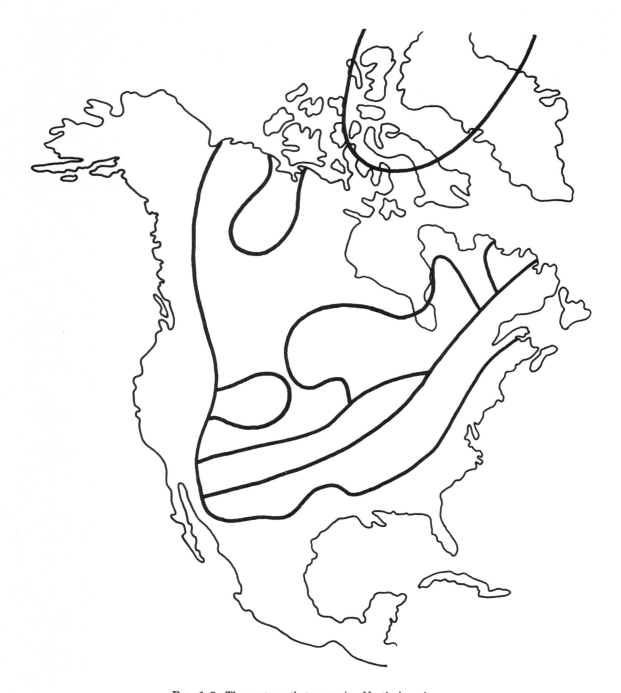

FIG. 1-9. The cratons that comprise North America.

1-10), and the sutures joining the landmasses are still visible as cores of ancient mountains called *orogens*. Caught between the cratons was an assortment of debris swept up by the drifting continents, including sediments from continental shelves and the ocean floor, stringers of volcanic rock, and small scraps of continents, all sliced up by faults. *Ophiolites*, which are pieces of ocean crust thrust up on land, have been dated as old as 3.6 billion years and are mined extensively for metal ores around the world. *Blueschists*, which are metamorphosed rocks of subducted ocean crust shoved up on the continents, also might have been present at such an early age.

Eventually, all the protocontinents came together into a single large supercontinent. Volcanic activity, magmatic intrusions, and rifting and patching of the continent continually built up the interior, while erosion and sedimentation built up the continental margins. Sedimentation rates were probably much higher than they are today, as a result of more vigorous weathering of rocks by frequent violent storms and lack of vegetative cover to hold the soil in place (since land plants were not yet in existence).

Presently, the continental crust is thickest under the Himalayan Mountains, where it is about 45 miles thick, but it averages 25 to 30 miles thick. The oceanic crust is much thinner; in most places, it is only 3 to 5 miles thick. The continental crust is twenty times older than the oceanic crust, which is found nowhere to be older than 200 million years.

Like rafts on a sea of molten rock, a dozen or so lithospheric plates (FIG. 1-11) carry the crust. The plates spread apart at midocean ridges, and they converge and subduct into the mantle at deep ocean trenches. The plates and oceanic crust are constantly being recycled through the mantle, but the continental crust always remains on the surface. The Moon and the other terrestrial planets have thick,

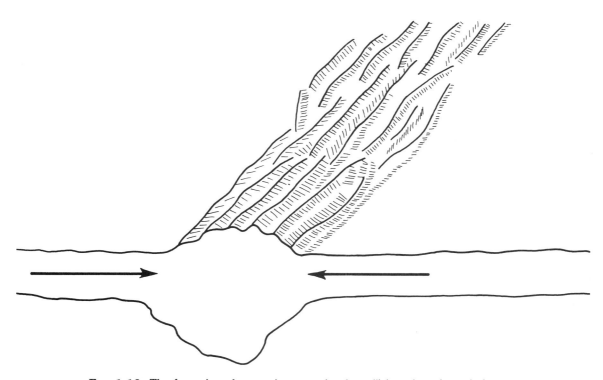

FIG. 1-10. The formation of mountain ranges by the collision of continental plates.

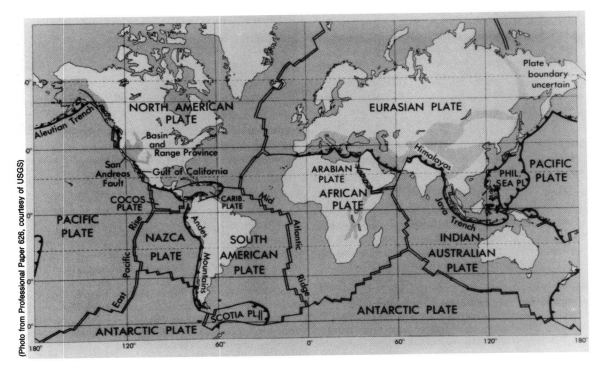

FIG. 1-11. The Earth's lithospheric plates.

buoyant, nonsubducting crusts because they are either too cold or too hot and thus have been tectonically dead for over 2 billion years. It is because of plate tectonics that life was able to evolve on Earth. It has even been suggested that there is active plate tectonics because the Earth contains life. Lime-secreting organisms in the ocean act as thermostats by removing carbon dioxide, an important greenhouse gas, from the atmosphere and storing it in ocean-bottom sediments, thereby keeping the Earth's surface within the temperature range in which plate tectonics can operate, which in turn keeps the Earth alive.

2

The Carbon Connection

THE character of Earth (FIG. 2-1) had a lot to do with its ability to generate and hold onto life for as long as it did. The large size of Earth—the largest of the terrestrial planets—keeps it from losing its interior heat so that it remains geologically active. Earth's density—the greatest of all the planets—provided enough gravity to hold onto a substantial atmosphere and a large ocean. Its distance from the Sun, 93 million miles ± 1.5 million miles, kept the Earth at an average temperature between the freezing and boiling points of water, within which life can exist. The tilt of its axis between 22 and 25 degrees gave Earth its seasons, which aided in the development of a huge variety of species. The high rotation rate gave Earth light and dark, hot and cold—an important condition in the early history of life. The nearby Moon gave Earth regular tides, which is thought to be a prerequisite for the initiation of life. Earth developed a thick atmosphere, which was reducing at first and became oxidizing when life became prolific.

Earth has more liquid water than any other planet. Water covers 70 percent of the Earth's surface, and 60 percent of Earth's surface is covered by water over 1 mile deep. Water has some unique properties. It has a very high specific heat, or *heat capacity*, and is an excellent solvent, able to dissolve most substances on Earth. There is so much water that if all the land were pushed off into the ocean, the water would cover the entire surface of the planet to a depth of up to 2 miles.

Earth also has a great deal of carbon. Much of the carbon came from the interior of the Earth in the form of carbon dioxide and other carbon compounds that erupted from volcanoes. Some carbon came from meteors like tons of coal falling out of the heavens.

Life on Earth is *carbon-based*, meaning that chains of carbon, hydrogen, and nitrogen make up every form of living matter, from the simplest amino acid to complex RNA (ribonucleic acid) and DNA (deoxyribonucleic acid). Life could just as easily have been based on silica, which is the second most abundant element on Earth and is a major component of rocks. However, carbon happens to be the element that life on Earth uses because of its relative abun-

(Courtesy of NASA)

FIG. 2-1. View of the Earth from Apollo 15.

dance and suitability for life. It would also be compatible with other life forms that might exist in the cosmos. Like the laws of physics, which are expected to work the same everywhere in the universe, the laws of organic chemistry probably work just as well. The genetic code in life on Earth could reflect a universal chemistry, and if there is life elsewhere in the universe, chemically speaking it should be very similar. The linking of carbon chains into life's building blocks might, therefore, have been inevitable from the start, given the chemistry of Earth's primordial soup.

THE SECOND ATMOSPHERE

Today, volcanoes erupt a variety of products, including water vapor, nitrogen, carbon dioxide, methane, ammonia, and sulfur dioxide (FIG. 2-2).

Water and carbon dioxide are particularly abundant in magma to help it flow easily. It is not unexpected that, early in the Earth's history, volcanoes erupted in much the same manner as they do today, only on a greater scale and more violently because of the higher temperature of the Earth's interior and larger amounts of volatiles, which made eruptions more highly explosive.

The early volcanoes were giants by today's standards, and like huge cannons, they shot rock fragments and ash 100 miles or more into space. Because the Earth had no atmosphere, the debris was not scattered by air currents but fell back around the volcanic vent, building up volcanoes to prodigious heights. Fountains of lava (FIG. 2-3) burst through cracks in the thin crust, and the Earth was paved over with basalt, forming *maria*, or basalt plains,

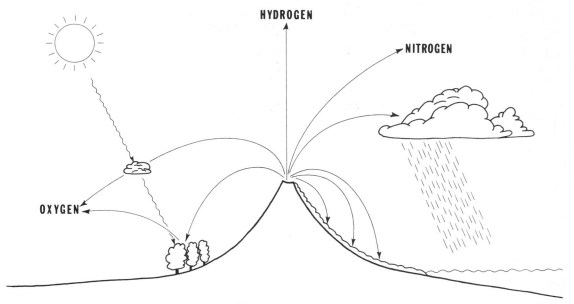

FIG. 2-2. The contribution of volcanoes.

(Photo by Garrett A. Smathers, courtesy of National Park Service)

FIG. 2-3. An Alae eruption of a Hawaiian volcano.

similar to those found on the Moon and Mars today. The entire surface of Earth was dotted by active volcanoes, which erupted one after another like thousands of belching industrial smokestacks.

Icy visitors from outer space pounded the Earth and provided a substantial amount of water vapor and gases. The barrage of meteors started around 4.2 billion years ago and lasted about 300 million years. Some meteorites were composed of rock, some of metal, and others of water ice and frozen gas. These meteorites along with comets, which are essentially rocks encased in ice, came from the outer reaches of the Solar System, where they formed from gases blown outward by the early solar wind. Thus, some of the volatiles that were lost when the original atmosphere blew away might have returned by way of comets and meteors. As the atmospheric pressure continued to rise, the smaller meteors burned up as a result of friction on entry into the upper atmosphere.

A great deal of carbon fell out of the sky from primitive forms of meteors called *carbonaceous chondrites*, which are chunks of carbon-rich rock with small spherical mineral inclusions, or *chondrules*, and are leftovers from the formation of the Solar System. Amino acids and DNA bases also are found in meteorites that formed elsewhere in the Solar System.

With all this heavy volcanic and meteoric activity going on, it did not take Earth long to develop a second atmosphere. Water vapor was so thick that the atmospheric pressure was several times greater than it is today. The surface was still very hot, and carbon dioxide, methane, ammonia (which broke down into nitrogen and hydrogen), and water vapor produced a *greenhouse effect*, which is the quality of a greenhouse that keeps plants warm in winter. The greenhouse gases kept the temperature of the early atmosphere well above the boiling point of water, even though the Sun shone only as feebly as it now does on Mars. The Sun was so weak that if Earth had today's atmosphere, the entire surface temperature would be the same as Antarctica in the dead of winter, and the ocean would be a solid block of ice. The greenhouse gases allowed sunlight to pass through them, but did not let the heat on the

Earth's surface escape into space, creating a runaway greenhouse effect.

A close comparison of the early Earth can be made with Venus, which has a heavy atmosphere composed almost entirely of carbon dioxide, along with thick layers of acid clouds. The rocks on Venus scorch at the temperature of molten metal, and hot acid rain immediately evaporates when it strikes the ground.

The early atmosphere of Earth did not possess significant amounts of free oxygen. Any oxygen that was produced from the breakdown of water vapor by ultraviolet light, called *photo dissociation*, quickly combined with metals on the surface and formed metal oxides, while the hydrogen escaped into space or combined with carbon. Oxygen also recombined with hydrogen and carbon monoxide to form water vapor and carbon dioxide. Some of the oxygen also could have reached into the upper atmosphere and created a thin ozone screen, which would have reduced the dissociation of water by ultraviolet light, thus preventing the loss of most of the Earth's water while keeping the generation of oxygen to a minimum.

In the absence of oxygen, the prebiotic atmosphere became *anaerobic*, or reducing. Chemical reactions between carbon and hydrogen, for instance, required that hydrogen give up an electron to carbon in order for the two elements to bond together. The presence of oxygen would interfere with this reaction because oxygen would preferentially react with the carbon, or oxidize it, to form carbon dioxide. (This type of reaction occurs when fossil fuels such as coal or petroleum are burned.) Therefore, life could not have begun in an atmosphere initially filled with oxygen because organic molecules cannot form in the presence of this gas.

THE ORGANIC SOUP

Before the Earth cooled enough to allow the condensation of water vapor, the atmosphere held about as much water as would later fill the entire ocean, whereas today, only about 1 percent of the water is in the atmosphere. In effect, steam made up a substantial portion of the early atmosphere. The steam reacted with sulfur dioxide and nitrous oxide from volcanic eruptions to form sulfuric acid and nitric acid. Acid rain precipitated, fell to the earth, and evaporated immediately upon hitting the hot ground. The water vapor was lofted high into the upper atmosphere, where it gave up its acquired heat to space and precipitated again, repeating the process in one continuous cycle. The process worked similarly to an air conditioner, which removes heat from the inside of a house and pumps it outside.

After millions of years, the precipitation-evaporation cycle cooled the surface sufficiently for rain to collect in small standing bodies of water. Earth was still quite hot so the evaporation rates were very high, and only a small fraction of today's ocean remained liquid on the surface. As Earth's temperature continued to drop, large amounts of water collected in meteor craters and collapsed volcanic craters called *calderas* (FIG. 2-4), which filled and overflowed onto flat lava plains. Volcanic peaks and jagged meteor craters soon were worn down by ero-

sion, and the high ground was plained off into a flat, low-lying surface called a *peneplain*.

The acidic rainwater reacted chemically with metallic minerals on the small continental masses, producing metallic salts that were carried in solution by streams emptying into the oceans. Rainwater also percolated into the ground, dissolved minerals from porous rocks, and was transported by aquifers out to sea. As the primitive ocean basins began to fill, seawater acquired a considerable salt content, along with other chemical substances. Solid rock exposed on the land was chemically broken down into clays and carbonates, and mechanically broken down into silts, sands, and gravels. Streams heavily laden with sediments often filled their beds, forcing them to take several detours as they meandered their way to the sea. When the streams reached the ocean, their velocity fell off sharply and their sediment load dropped out of suspension, while chemical solutions became thoroughly mixed with seawater through currents and wave action.

(Courtesy of USGS)

FIG. 2-4. Crater Lake, Oregon.

The sediments continually built the continental margins outward, with coarser sediments accumulating near shore and progressively finer sediments settling farther out to sea. As the shoreline moved seaward, the original fine sediments were covered by coarser sediments. As the shoreline receded because of higher sea levels, coarse sediments were overlain by fine sediments, giving the sedimentary sequence a layered structure. Also, carbonates chemically precipitated and accumulated in thick layers on the ocean floor.

As the weight of the overlying sedimentary layers pressed downward on the lower strata, the sediments were lithified into solid rock, providing a geologic column of alternating layers of limestone, shales, siltstones, and sandstones. If these rocks were subjected to the heat and pressure of Earth's interior, they were metamorphosed into marble, slate, quartzite, and schist respectively. Metamorphosed sediments are among the oldest rocks known, suggesting that a weathering cycle was already in place some 3.8 billion years ago. It is also quite possible that life is even older than these rocks.

The chemical evolution of the ocean would not be complete without a primitive form of plate tectonics operating at an early stage. The original ocean crust was composed of basalt lava flows, which erupted on the surface before the ocean basins began to fill. Embedded in the thin crust were granitic blocks, or "rockbergs", that assembled into cratons, or microcontinents. The cratons were lighter than basalt and remained on the surface, drifting freely as convection currents in the mantle pushed or pulled them along.

The small continental masses did not plow through the ocean crust like an ice breaker plows through Arctic ice, but instead were carried along on lithospheric plates. When a continental plate pressed against an oceanic plate, the latter bent downward and subducted under the continental plate. The oceanic plate remelted in the Earth's interior, acquired new minerals from the mantle, and reemerged at volcanic centers called *spreading ridges*, which rejuvenated the ocean crust and spread it wider apart (FIG. 2-5). Sediments deposited on the ocean floor, along with water trapped between sediment grains, also were caught in the subduction zones, but because of their low melting points and light weight, they were buoyed to the surface to supply volcanoes with magma and seawater that was recycled through the magma. This water was called *juvenile water*. Seawater also

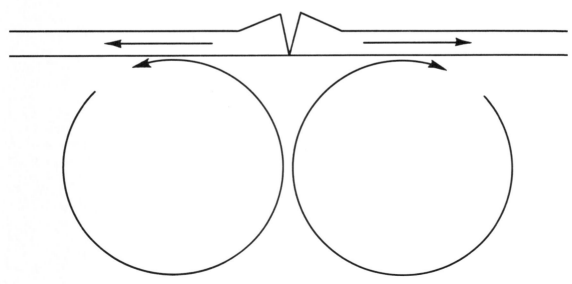

FIG. 2-5. Convection currents in the mantle spread lithospheric plates apart.

seeped into cracks in the ocean crust, where it was heated near magma chambers, leached water-soluble compounds from the rock, and rose to the ocean floor, forming undersea hydrothermal vents.

This interchange between the crust and the mantle (FIG. 2-6) constantly resupplied the ocean with chemical substances, some of which precipitated on the ocean floor. At the same time, surface runoff carried substantial amounts of weathered materials from the continents and deposited them into the ocean. Weather systems were much more vigorous than they are today as a result of the warm climate. Therefore, sedimentary rates were much higher, although the size of the continents were smaller, so that the overall input of sediments into the sea was probably not much different than it is today.

The ocean became a chemical dumping ground, and instead of a slow process of accumulating salts and other chemical substances over a lengthy period of time, it achieved chemical equilibrium early in its history. Thus, the ocean became a vast chemical factory, synthesizing a variety of inorganic compounds, and the stage was set for the appearance of life much sooner than it was ever before imagined.

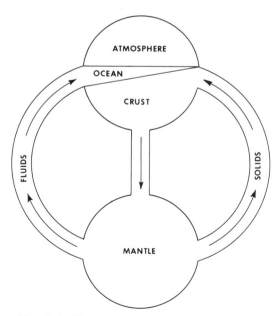

FIG. 2-6. The steady-state model of the Earth.

THE SPARK OF LIFE

All 20 amino acids, the very building blocks of life that make up proteins and DNA, have been artificially produced by spark discharge experiments using a specially designed apparatus and a variety of ingredients that were thought to comprise the primordial sea. The first of these spark chambers was designed in 1953 by the American chemist Stanley Miller while he was a graduate student at the University of Chicago.

As part of his doctoral thesis, Miller, who worked under the Nobel Prize-winning chemist Harold Urey, attempted to synthesize amino acids by recreating conditions that existed on the early Earth. A flask of boiling water, representing the ancient seas, was connected to a large spherical vessel above it containing ammonia, nitrogen, methane, and water vapor, representing the early atmosphere. Inside the vessel were two electrodes. Through them were passed 60,000 volts of electricity to simulate lightning and strong ultraviolet radiation from the Sun. After passing through the spark gap, the gases were condensed and collected below in a trap with a return line back to the boiling flask. Thus, gases and water vapor circulated in a closed-loop system, representing the hydrologic cycle.

After about a week, a dark soup was collected from the trap and analyzed. It was found to contain an assortment of carbon compounds, among which were amino acids. This experiment proved that life was not a quirk of nature, but followed certain logical rules by which chemistry operates. Therefore, given the conditions of the early Earth, life would have been all but inevitable.

The warm climate of the early Earth supported the evaporation of large quantities of water from the ocean, creating a powerful atmospheric heat engine. Huge billowing clouds shrouded the Earth and raced around the globe, carried along by energetic air currents. An unbalanced charge distribution among clouds and between clouds and the ground created powerful lightning bolts that leaped across the sky in a continuous display of sparks (FIG. 2-7). Upon striking the ground, lightning bolts dissipated their energy with a mighty blast whose tremendous heat fused rocks. Cauliflower-shaped eruption clouds

FIG. 2-7. Large lightning bolt.

(Courtesy of NOAA)

from numerous gigantic volcanoes produced lightning, which darted across the ash cloud. In addition, strong ultraviolet radiation from the Sun penetrated the atmosphere.

These two energy sources were able to catalyze chemical reactions in the primitive atmosphere. Organic molecules formed and were rained out by heavy showers. The water returned to the ocean in streams, carrying with it carbon compounds, including those formed in the atmosphere and those leached out of the rock. In the ocean, chemical reactions among carbon compounds produced a huge variety of hydrocarbon chains.

The carbon chains strung together to form molecules of hydrogen cyanide, ethane, ethylene, and formaldehyde, some of Earth's first steps toward life. Formaldehyde molecules were of particular importance because they joined to form sugars such as *ribose*, a constituent of RNA that is a conveyor of the genetic code in DNA.

The pattern specified in DNA's genetic code that is used to synthesize proteins is copied from the DNA template onto RNA, which becomes the blueprint for building proteins. Amino acids take their proper places along the RNA strand according to a series of linked three-letter words called *codons*.

RNA's ability to catalyze certain protein reactions and to replicate itself in the absence of proteins strongly supports the hypothesis that primordial RNA could have been the original genetic material. Indeed, RNA has been shown to evolve in a test tube. Within a short time after the forma-

24 THE CARBON CONNECTION

tion of the ocean, all the essential amino acids and nucleotides—the subunits of RNA and DNA—were present, and the ocean became a rich broth of organic compounds.

LIFE'S FIRST STEPS

No matter how varied life forms are on Earth today, from the simplest bacteria to man, their central molecular machinery is exactly the same. Every cell of every organism is constructed from complex protein molecules composed of the same set of 20 amino acids. All organisms use ATP (adenosine tri-phosphate) to transfer energy from one part of the cell to another. Strands of DNA are built into a left-handed double helix—a sort of molecular spiral staircase—and never has there been found a right-handed double helix.

Whether it is plant or animal, the operation of the genetic code in protein synthesis always works the same. The system is practically perfect. If one protein molecule is out of place, a mutation occurs that might be beneficial or detrimental. Cells that construct tissues, which in turn make up organs, all must be just right or the organ might fail to work properly. With so much similarity in living beings, it seems reasonable to conclude that all life sprang from a common ancestor, and that all alien forms of life, of which there are no present-day descendants, became extinct early in the history of life on Earth.

Since life made its appearance within the first billion years of Earth's existence, it had to evolve into complex organisms out of simple materials rather quickly. Laboratory experiments have synthesized almost all of the components of life, but have yet to put them all together into a living cell. Extraterrestrial influences can be ruled out, because life appears to be of this world, even though organic molecules have been found in 4.5 billion-year-old meteorites. If given enough time—a couple of billion years or so—random combinations and permutations could give rise to an entity that has the fundamental properties of life: the abilities to store energy, catalyze chemical reactions, and self-replicate. Clay has all of these qualities and has been around since the very beginning, although it is not considered to be alive in the normal sense.

Cycles of wetting and drying in clays, caused by the ocean tides (FIG. 2-8), can link molecules of amino acids and transfer energy from the environment to the organic molecules. The ions in clay also can exert selective catalytic effects on amino acids, and some organic molecules can perform enzyme-like functions in association with clays.

Clay crystals replicate by means of spontaneous crystallization, in which seed crystals, or *crystal genes*, provide the template on which silicon atoms and metal ions can grow, layer by layer. The crystal gene grows sideways, spreading its message down lengths of folded or branched membranes or flakes of constant thickness. *Mutation*, a defect in a genetic crystal, would control the growth of interweaving crystals, possibly making the clays more suitable to their particular environment. Thus, in the inorganic world a sort of natural selection operated to produce the best species of clay for a particular environment.

In this early world of rock (FIG. 2-9) organic molecules acted as catalysts for clay synthesis. Some amino acids can make metal ions, such as aluminum, more soluble for clays. Metal ions, such as zinc and copper, can act as go-betweens to bind nucleotides to clays. Other types of organic molecules stick to clays, often altering their physical properties. Organic molecules can change the shape and size of clay crystals, thereby affecting their growth. Organic polymers could act as a glue to hold clay particles together.

The precursors of RNA could be used as structural materials for clays. The negatively charged spine of RNA would stick to the edges of positively charged clay particles and perhaps "read" the information in clay genes. It then could act as a carrier of genetic information along the entire length of clay strands. Genetic takeover would occur when RNA acquired the ability to self-replicate. It then could carry the clay's genetic code elsewhere and influence the growth of other clays. Eventually, RNA would begin to build intricate structures, not out of inorganic molecules in clay, but out of organic molecules that were much better materials to work with, and the first proteins would be synthesized.

Proteins by themselves are not living matter.

FIG. 2-8. The creation of the tides. A is the common center of gravity between the Earth and the Moon, which causes the ocean to bulge at two ends.

FIG. 2-9. Boiling mud springs northwest of Imperial Junction, California.

However, when RNA began to manufacture proteins instead of clay, it took a major step up the evolutionary ladder. Silicic acid was replaced by amino acids as a building block for simple proto-organisms that were not yet alive, but existed on the borderline of life and nonlife. With more complexity and the possible inclusion of DNA, prebiotic organisms crossed over the threshold of life. It was life of the lowest order from which all future life forms would spring.

The first living organisms were extremely small noncellular blobs of protoplasm called *prokaryotes*, which did not have a nucleus. Reproduction was asexual by simple fission, whereby the organism either split in two or small parts of the organism budded off and began growing independently, providing little or no variety among species. There was plenty of food available, and the self-duplicating organisms lived on an abundance of organic molecules in the primordial sea. As a result, a rapid chain reaction was set off, resulting in phenomenal growth. The organisms had no means of locomotion, but drifted freely in the ocean currents until they were dispersed throughout the world.

Although the first simple organisms appeared to have arrived fairly soon after conditions on Earth became favorable, it took almost another billion years before anything resembling present-day life forms came along.

THE BIOSPHERE

Now that the Earth contained life, slow but steady changes were made that greatly affected the final outcome of the planet. Like the Earth, other planets and satellites in the Solar System have a core, a mantle, a crust, and even an atmosphere or an icy hydrosphere, but none of them have a biosphere. Just having living entities is not enough, either. Life must be integrated with the geosphere, hydrosphere, and atmosphere to constitute the biosphere. Biologists have cataloged some 1.5 million species of animals and 0.5 million species of plants. Because many species have evaded detection, the total number could easily be ten times as many, which indicates the quantity and diversity of life on Earth.

Since life's first humble beginnings, it has responded to a variety of chemical, climatological, and geographical changes in the Earth, forcing species either to adapt or perish. Many dead-end streets along branches of the evolutionary tree are found in the fossil record, which itself only represents a fraction of the species that have actually lived. Just about every conceivable form and function have been tried, some with more success than others. It is through this trial-and-error method of specialization that natural selection has chosen some species to prosper, while condemning others to extinction.

Very few places on Earth are truly devoid of life. Life is found in the hottest deserts and the coldest polar regions. It resides in the lowest canyons and the tallest mountains. Life also exists in the deepest oceans and the highest regions of the troposphere. Nor is life excluded from scalding hot springs or areas deep below the ground.

Although species most frequently encountered on the Earth's surface would seem to be the most dominant force in shaping the Earth, it is actually the unseen microbes, which constitute about 90 percent of the biomass, that pull the most weight. They are morphologically simple creatures, but are bi-ochemically diverse and highly adaptive. Eighty percent of the Earth's breathable oxygen is generated by photosynthetic single-cell organisms in the ocean.

Microorganisms like bacteria play a critical role in breaking down the remains of plants and animals so they can be recycled. Surface plants depend on bacteria in their root systems for nitrogen fixation. Bacteria live symbiotically in the gut of animals, even humans, and aid in the digestion of food. Biological processes are responsible for massive concentrations of silicon, carbon, iron, manganese, copper, and sulfur in the Earth's crust. Simple organisms also comprise the bottom of the food chain on which life ultimately depends for its survival.

Life on Earth constitutes a geological force that is totally lacking on the other planets of our Solar System. The evidence of biospheric processes in the Earth's history belongs to the broad field of biogeology, and what appears to be the earliest fossilized remains of microorganisms go as far back as 3.5 billion years. The 3.8 billion-year-old carbonaceous sediments of the Isua formation in southwest Greenland show a depletion of carbon-13 with respect to carbon-12, which is thought to be a common manifestation of biological activity. Therefore, life processes might have operated on the Earth for at least four-fifths of its geologic history (FIG. 2-10).

With this much time involved, it is not surprising that life brought about some dramatic and far-reaching changes to the Earth. The first major alteration came with the deposit of banded iron formations on continental margins by iron-eating bacteria. These formations currently are mined extensively for iron ore around the world. The second major change was the conversion of most of the carbon dioxide in the ocean and atmosphere into oxygen when photosynthesis as an energy source evolved. The oxygen produced a secondary benefit, the ozone layer in the upper stratosphere, which made conditions safe for land plants to cover the Earth, (which also constituted a major change).

Living things can store energy by combining carbon from carbon dioxide, for example, with hydrogen from water into hydrocarbons. Thick coal deposits, which are essentially buried solar energy because they originated as vegetative matter, and

(Photo from Earthquake Information Bulletin, courtesy of USGS)

FIG. 2-10. Geologic time spiral.

vast subterranean reservoirs of oil, which are cooked hydrocarbon molecules from once living microorganisms, have been accumulating over eons. When fossil fuels are burned in automobiles and factories, the equation is reversed, and carbon is recombined with oxygen, releasing energy and carbon dioxide.

The burning of tremendous amounts of fossil fuels, the pollution of the environment with toxic wastes, the destruction of forests and wildlife habitat, and the uncontrollable human population explosion places man in the unique position of causing major changes to the Earth in a comparatively short period of time. This makes man a major biogeological force on the face of the Earth, and it remains to be seen what kind of world will result from his activities.

3

Age of Early Life

THE first 4 billion years, or about 90 percent of the Earth's history, is known as the *Precambrian era*, the longest and least understood period of Earth. The Precambrian is subdivided into the Archean eon, or time of initial life, from 4.6 to 2.5 billion years ago, and the Proterozoic eon, or time of earliest life, from 2.5 to 0.6 billion years ago. The boundary between the Archean and Proterozoic eons is somewhat arbitrary and reflects major differences in the character of rocks older than 2.5 billion years and those of a younger age.

The Archean was a time when the Earth's interior was hotter, the crust was thinner and more unstable, and the tectonic plates were more mobile. The Earth was still subjected to a high degree of volcanism and periodic meteor bombardment. The Proterozoic was a shift to calmer times as the Earth matured from adolescence into adulthood. Continents became more stable and were welded into one large amalgamation or supercontinent.

Marine life during the Proterozoic was distinct from that of the Archean and represented a considerable advancement of species. The global climate was cooler in the Proterozoic than in the Archean, and the Earth experienced its first major ice age about 2 billion years ago.

The Proterozoic came to a close around 600 million years ago after a second major ice age. An explosion of species, representing nearly every major group of marine life, set the stage for the Phanerozoic eon (time of later life), and for the first time, fossilized remains became prevalent.

THE ARCHEAN EON

The Archean eon is often depicted as an unsettled time for the Earth. During this interval, the Earth experienced its worst growing pains, and this restlessness might have been a major contributing factor to the emergence of life. The first 700 million years of Archean time—also called the *Hadean eon*, from the Greek word for "hell"—is missing from the geologic record because no rocks could survive the tumultuous activity that took place on the Earth's surface during that eon. Heavy turbulence in the mantle, with a heat flow three times what it is today, produced violent agitation on the surface, resulting in a sea of molten and semimolten rock broken up by giant fissures, from which fountains of lava spewed skyward. Jangled pieces of solid crust jostled against each other, broke up, and slid under the magma like ships sinking on the high seas. Meteors up to 60 miles wide added to the mayhem as they hammered Earth from above, supplying their own special ingredients to the cauldron.

According to one theory, during the Hadean eon a large asteroid was kicked out of its orbit in the asteroid belt lying between Mars and Jupiter by the gravitational pull of Jupiter. The asteroid slammed into Earth, and the debris from the collision coalesced into the Moon. The closeness of the Moon caused a huge tidal bulge, which further mangled the Earth's surface. When the Earth acquired an ocean, the nearby Moon caused a great sloshing of water back and forth within the basin.

When Earth was about 0.5 billion years old, most of the short-lived, energetic radioactive elements in the mantle decayed into stable daughter products, and the mantle gradually began to cool. This marked the beginning of a more or less permanent crust composed of a thin layer of basalt embedded with scattered blocks of granite. The granite combined into stable bodies of basement rock upon which all other rocks were deposited (FIG. 3-1 THROUGH 3-3). The basement rocks formed the nuclei of the continents and are presently exposed in broad, low-lying, domelike structures called *shields* (FIG. 3-4).

Greenstone belts, which are dispersed among or around the shields, are a jumble of metamorphosed lava flows and sediments, possibly from *volcanic arcs* (volcanic islands on the edges of subduction zones) caught in the squeeze between continents. The rocks, which are greenish because of the presence of the mineral chlorite, might be evidence that plate tectonics operated early in the Archean. *Ophiolites*, which are slices of ocean floor shoved up on the continents by drifting plates and are dated as old as 3.6 billion years, were also caught in the greenstone belts. A number of ophiolites contain ore-bearing rocks, which are important mineral resources. Because greenstone belts are essentially Archean in

age, their disappearance around 2.5 billion years ago marks the end of the Archean eon.

Archean rocks are important sources of ore deposits the world over. The mineralized veins either are Archean in age or they invaded Archean rocks at a much later date. The ore deposits of Archean age are remarkably similar the world over. Gold of Archean age is mined on every continent except Antarctica. In Africa, the best gold deposits are found in rocks as old as 3.4 billion years. In North America, the best gold mines are in the Great Slave region of northwest Canada, where there are over 1000 known deposits. Most of the gold deposits are found in greenstone belts invaded by hot magmatic solutions from the intrusion of granite bodies into the greenstones, and the gold occurs in veins associated with quartz.

Iron is also widespread in Archean rocks and is found on all continents, but is not nearly as important as iron of Proterozoic age. Half the world's production of chromium comes from Archean rocks of South Africa. The world's greatest nickel deposit at Sudbury, Canada, is in Archean rocks. The great copper belt of Zambia, which is estimated to contain a quarter of the world's copper, is of Archean age. Most diamonds are derived from younger volcanic intrusives that cut Archean rocks.

Life in the Archean eon consisted mostly of bacteria, unicellular or noncellular algae, and clusters of algae in limestone called *stromatolites* (FIG. 3-5). The cells lacked a distinct nucleus and therefore are called *prokaryotes*, from the Greek *karyon*, meaning "nutshell." They lived under anaerobic (lacking oxygen) conditions and depended for the most part of extracellular sources for nutrients. These sources consisted of a rich supply of organic molecules that were constantly being generated in the sea around them. Most organisms used a primitive form of metabolism called *fermentation* to convert nutrients into energy (FIG. 3-6). Fermentation is a less efficient means of metabolism,and releases energy by breaking down simple sugars such as glucose into smaller molecules by enzymes.

Primitive types of photosynthesis probably began about 3.5 billion years ago, but the organisms that initiated the process were bacteria, and like

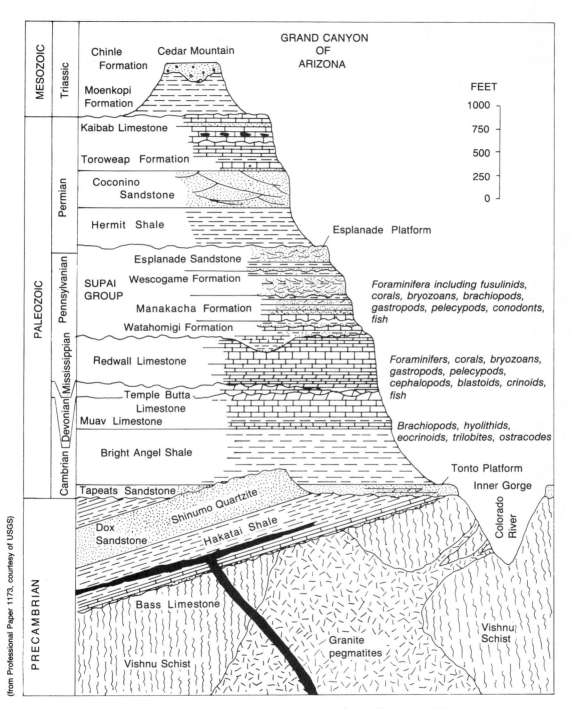

FIG. 3-1. Geologic cross-section of the Grand Canyon of Arizona.

FIG. 3-2. View of the Grand Canyon from Toroweap Point, Arizona.

those of today were best suited for an oxygen-poor environment. Therefore, oxygen, which was poisonous to primitive life forms, was kept at very low levels by its combination with dissolved metals in seawater and reduced gases from submarine hydrothermal vents.

The oldest evidence of life are *microfossils*, which are the remains of ancient microorganisms, and stromatolites which are layered structures formed by the accretion of fine sediment grains by matted colonies of cyanobacteria, or blue-green algae. They were found in 3.5 billion-year-old sedimentary rocks of the Warrawoona group in western Australia at an isolated place called North

FIG. 3-3. The Precambrian Vishnu Schist, Grand Canyon National Park.

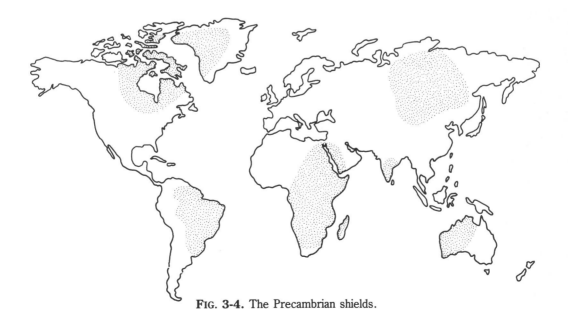

FIG. 3-4. The Precambrian shields.

FIG. 3-5. Stromatolite beds, Gila County, Arizona.

Pole. Associated with these rocks were *cherts*, which are rocks composed of microscopic grains of silica, with *microfilaments*, which are small, thread-like structures of possible bacterial origin. Most Precambrian cherts are thought to be chemical sediments precipitated from silica-rich water in deep oceans.

The abundance of cherts in the Archean eon could be used as evidence that most of the Earth's crust was deeply submerged during that time; however, cherts at North Pole appear to have had their origin in shallow water. Silica was leached out of volcanic rocks that erupted into shallow seas (FIG. 3-7). The silica-rich water circulated through porous sediments, dissolving the original minerals and precipitating silica in their place. Microorganisms buried in the sediments were thus encased in one of the hardest natural substances on Earth, and the microfossils

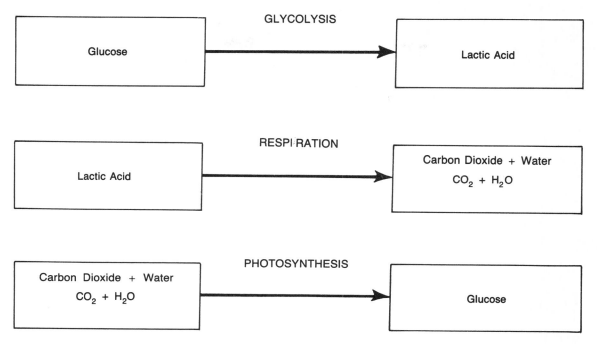

Fig. 3-6. Fundamental mechanisms of energy extraction and storage.

were able to survive the rigors of time.

Similar cherts with microfossils of filamentous bacteria dated 3.2 to 3.3 billion years old were found in eastern Transvaal, South Africa, and cherts 2 billion years old were found in the Gunflint iron formation on the north shore of Lake Superior. The rocks near Lake Superior originally were mined for flint, which was used to fire the muzzle loaders of the early settlers until the discovery of iron.

Stromatolites, from Greek meaning *stony carpet*, are a less direct evidence of life because they are remains, not of the microorganisms themselves, but of the sedimentary structure the microorganisms built. The stromatolites of North Pole, Antarctica are distinctly layered accumulations of calcium carbonate with a rounded, cabbagelike appearance. Modern stromatolites are very similar and are concentrically layered mounds of calcium carbonate built up by bacteria or algae, which cement sediment grains by secreting a jellylike ooze.

The size and shape of the Archean-age microfossils and the form of the stromatolites they built indicate that the organisms were either oxygen-releasing or sulfur-oxidizing photosynthetic organisms that were dependent on sunlight for their growth. Just as stromatolites do today, Precambrian stromatolite colonies lived in the intertidal zones. Their height, which was more than 30 feet tall in some cases, was indicative of the height of the tides. The oldest colonies were much taller than modern varieties and suggest that the Moon was much closer to Earth, and therefore raised higher tides.

THE PROTEROZOIC EON

By the beginning of the Proterozoic eon 2.5 to .6 billion years ago, most of the material presently locked up in sedimentary rocks was already at or near the surface, and there were ample sources of Archean rocks available for erosion and redeposition. Sediments derived directly from primary sources are called *wackes*, and most Proterozoic wackes composed of sandstones and siltstones were derived from Archean greenstones.

Another common rock in the Proterozoic was fine-grained quartzite, which is derived from the erosion of siliceous grainy rocks such as granite and *arkose*, a coarse-grained sandstone. *Conglomerates*, which are consolidated equivalents of gravels, were also abundant in the Proterozoic. Nearly 20,000 feet of Proterozoic sediments are found in the Uinta Range of Utah (FIG. 3-8), and the Montana Proterozoic belt system has sediments over 11 miles thick.

The Proterozoic eon is also known for its terrestrial red beds, so named because grains of sand were stained red by iron oxide. Their appearance about 1 billion years ago is an indication that the atmosphere had a substantial amount of oxygen. Extensive red bed deposits occur in the western United States (FIG. 3-9).

The weathering of primary rocks also produced solutions of calcium carbonate, magnesium carbonate, calcium sulfate, and sodium chloride, which in turn precipitated directly into limestone, dolomite, gypsum, and halite respectively. These are thought to be mainly chemical precipitates and not of biological origin. In the Mackenzie Mountains of northwest Canada are deposits of dolomite 6500 feet thick. Carbonate rocks such as limestone and chalk, produced chiefly by organic process involving shells and skeletons of simple organisms, became much more common during the Proterozoic eon between about 700 and 570 million years ago, than in the Archean, when they were relatively rare because of the scarcity of lime-secreting organisms.

(Photo by Robert T. Haugen, courtesy of National Park Service)

FIG. 3-7. Molten lava pouring into the sea.

(Photo by W. R. Hansen, courtesy of USGS)

FIG. 3-8. Red Castle Peak, Uinta Mountains, Summit County, Utah.

The continents in the Proterozoic eon were composed of odds and ends of Archean cratons. Four or five cratons were welded together to form what is now central Canada and northcentral United States. Continental collisions continued to add a large area of new crust to the growing protocontinent of North America. The better part of the continental crust underlying the United States from Arizona to the Great Lakes to Alabama formed in one great surge of crustal generation around 1.8 billion years ago. This surge has been unequaled in North America since, perhaps because plate tectonics and therefore crustal generation operated at a faster rate during the Proterozoic than it does today.

The presence of volcanic rock near the edge of North America implies that it was part of a super-continent in the early Proterozoic. Because the central part of the supercontinent was far removed from the cooling effects of subducting plates, it heated up and erupted with volcanism, and the warm, weakened crust broke apart. Toward the end of the Proterozoic around 600 million years ago, another supercontinent, located near the equator, broke apart into possibly four major continents, although they were not arranged in their present pattern. The breakup produced 12,000 miles of new continental margin and might have played a major role in the explosion of new organisms at the end of the Proterozoic.

Ore deposits of the Proterozoic are characterized by being bedded or stratified, rather than in veins as in the Archean. Of chief importance during the Proterozoic was iron, which is the fourth most abundant element in the Earth's crust. Iron, leached from the continents, was dissolved in seawater under reducing conditions. When photosynthetic organisms began producing oxygen in the ocean, it combined with iron in solution, keeping the oxygen level within tolerable limits for the early prokaryotes.

Throughout the Archean, the percentage of oxygen probably remained under one percent because of this regulating mechanism. Between 2.6 and 2 billion years ago, enough oxygen was produced to react with iron on a large scale. Iron was precipitated in vast amounts and deposited along with sediments on shallow continental margins. Alternating bands of iron-rich and iron-poor sediments gave the ore a banded appearance, thus prompting the name *banded iron formation* (BIF). In effect, biological activity was responsible for the iron deposits since the oxygen was produced by photosynthetic organisms. After most of the iron in seawater was locked up in sediments, the level of oxygen began to rise markedly.

BIF deposits are extensive (FIG. 3-10) and provide over 90 percent of the world's minable iron reserves. Biochemical activity in the ocean was also responsible for stratified sulfide deposits. Sulfur-eating bacteria living around undersea hydrothermal vents oxidized hydrogen sulfide into elemental sulfur and various sulfates. Copper, lead, and zinc, which were much more abundant in the Proterozoic than in the Archean, also reflect a submarine volcanic origin.

Proterozoic life was substantially more advanced and complex than Archean life. Organisms evolved very little over their first billion-year stay on Earth because of their asexual reproduction. This type of

(Photo by N. H. Darton, courtesy of USGS)

FIG. 3-9. Red bed formation on the east side of the Bighorn Mountains, Johnson County, Wyoming.

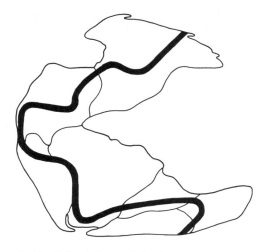

FIG. 3-10. Belt of Precambrian iron deposits running through Pangaea.

reproduction involved simple fission, which does not allow much chance for change, and a primitive form of metabolism, which kept the organisms in a low-energy state.

The first major advancement in the Proterozoic came with the development of an organized nucleus and sexual reproduction, which produced a new breed of single-celled organisms called *eukaryotes* around 1.5 billion years ago. The advantage of a nucleus is that chromosomes divide and unite hereditary material in a systematic way. This method allows genetic mutations to occur, which produces slightly different individuals, some of which might be more adaptive to their environment and are able to pass on their "good genes" to their offspring.

Single-celled eukaryotes are on average about ten times larger than prokaryotes, and their cells contain a variety of organells, which produce energy and carry out other vital functions while living symbiotically within their hosts. Metabolism in eukaryotes is by *respiration*, in which glucose is *oxidized*, or burned, in the presence of oxygen. Metabolism releases 16 times more energy than the more primitive fermentation. Therefore, the presence of eukaryotes indicates that the ocean had a substantial amount of oxygen at that time.

THE EARLIEST ANIMALS

At first, single-celled animals called *protists* (FIG. 3-11) shared some characteristics with plants, and there was no clear demarcation between the two. The cells contained elongated structures of *mitochondria*, bacterialike bodies that produce energy by oxidation, and chloroplasts, packets of chlorophyll that are like blue-green algae and provide energy by photosynthesis. Some cells moved about with a thrashing tail called a *flagellum*, which resembled a filamentous bacterium that joined up for mutual benefit. Other cells had tiny hairlike appendages called *cilia* that helped them get around by rhythmically beating the water. Still others, like the amoeba, traveled by extending fingerlike protrusions outward from the main body and flowing into them.

The ability to move about under their own power is what essentially separates animals from plants, although some animals perform this function only in the larval stage and become sedentary or fixed as an adult. It is this mobility that enabled the first animals to feed on plants and other animals, and this established a new predator-prey relationship.

Each tiny organism was like a committee of simpler ones (the *organells*) that were probably taken into the cell but, instead of being digested, survived to collaborate in a sort of communal life. Many protozoans secreted a tiny shell composed of calcite, and when the animals died their shells sank to the bottom of the ocean. Over time, these shells built up impressive formations of limestone (FIG. 3-12). When the cell divided sexually, the DNA in the nucleus as well as the DNA in the organells replicated, and half the genes stayed with the parent and the other half were passed on to the daughter cell. This increased the possibility of genetic variation, which greatly accelerated the rate of evolution as organisms encountered new environments.

Multicellular animals called *metazoans* evolved in the latter part of the Proterozoic eon around 750 million years ago, when the buildup of oxygen in the ocean reached about 7 percent of its present value. The first metazoans were a loose organization of individual cells united for common purposes such as locomotion, feeding, and protection.

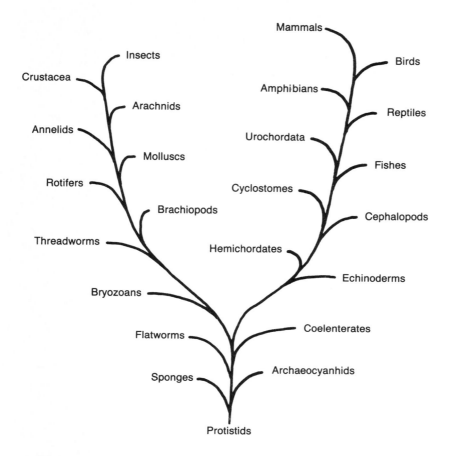

Crustacea
Insects
Arachnids
Annelids
Molluscs
Rotifers
Brachiopods
Threadworms
Hemichordates
Bryozoans
Flatworms
Coelenterates
Sponges
Archaeocyanhids
Protistids

Mammals
Birds
Amphibians
Reptiles
Urochordata
Fishes
Cyclostomes
Cephalopods
Echinoderms

FIG. 3-11. Family tree of animals.

The most primitive metazoans were probably composed of a large number of cells, each with its own flagellum. They were grouped into a small, hollow sphere, and their flagella beat the water in unison through some means of communication and propelled the tiny ball around. From these animals evolved sedentary creatures that were turned inside out and attached to the ocean floor. They had openings to the outside and the flagella, now on the inside, produced a flow of water, providing a crude circulation system for filtering food particles and ejecting wastes. These were the forerunners of the sponges, the first giants of the sea. Some species of sponges were probably ten feet or more across.

The next step up the ladder were the jellyfish, which had two layers of cells separated by a gelati-

nous substance, giving the saucerlike body a means of support. Unlike the sponges, the cells of the jellyfish were incapable of independent survival and were linked by a primitive nervous system, which caused them to contract, becoming the first simple muscles used for locomotion. Muscles and other rudimentary organs, including sense organs, along with a central nervous system evolved with the primitive segmented worms. These worms left behind a preponderance of fossilized tracks, trails, and burrows, so much so that the Proterozoic has often been described as the "age of worms."

By the time the Proterozoic eon came to a close and the Earth entered the Phanerozoic eon, the sea contained large populations of widespread and divers species. The dominant organisms were the coelenterates. Two species of this organism were jellyfish-like floaters up to 3 feet wide, and colonial feathery forms, probably ancestors of the corals, that were attached to the seafloor and grew more than a yard long. The rest of the organisms were mostly marine worms, unusual naked arthropodlike animals, and a curious-looking three-rayed, tiny naked starfish.

A sheetlike marine worm grew nearly three feet long but was less than one-tenth inch thick. This structure gave it a large surface area on which to absorb oxygen and nutrients directly from seawater. The unusual, flattened bodies of many animals were also the result of the limited supply of food available during the Precambrian. A high ratio of surface area to volume was needed to collect sunlight for algae, which lived in symbiosis within their host's body providing nutrients and removing waste products.

Many of the strange life forms were probably the result of adaptations to the highly unstable conditions that existed in the late Precambrian eon. As a consequence of overspecialization, many species never crossed the boundary that separated the Precambrian and the Cambrian ages. Those that did make it flourished in the Cambrian seas, but only a few of them gave rise to anything living today.

THE THIRD ATMOSPHERE

The appearance of blue-green algae, or its predecessor, a photosynthetic bacteria called *green sulfur bacteria*, around 3.5 billion years ago and the development of photosynthesis were possibly the most important steps in the history of life on Earth. Today they are still the most important in terms of maintaining the balance of nature. Primitive bacteria were limited by the amount of organic molecules being produced in the ocean. The first green-plant photosynthesizers, called *proalgae*, were probably intermediate between bacteria and blue-green algae and were able to switch from fermentation to photosynthesis and back again, depending on their environment. Because sunlight could only penetrate the ocean depths a couple hundred feet, the proalgae were confined to fairly shallow water.

FIG. 3-12. Eastern Sawtooth Range composed of Mississippian carbonates, Lewis and Clark County, Montana.

Tapping into the Sun provided an almost un-limited supply of energy, and the growth of photosynthetic organisms was phenomenal. The population explosion would have gotten out of hand if it were not for the fact that oxygen, which was generated as a waste product of photosynthesis, was also poisonous. If it were not for oxygen sinks, such as dissolved iron and other metals in the ocean, and the development of special enzymes to help organisms cope with and later use oxygen, life could have been in very serious trouble.

Photosynthesis uses energy from sunlight to split water molecules into molecular oxygen and elemental hydrogen, which then combines with carbon dioxide to form simple sugars by the formula: $H_2O + CO_2 = (CH_2O)n + O_2$ (FIG. 3-13). In order to create and maintain an oxygen-rich atmosphere, carbon dioxide used in the photosynthetic process had to be buried in the geologic column as carbonate rocks faster than oxygen was consumed by the oxidation of carbon, metals, and reduced volcanic gases.

About 2 billion years ago when the deposition of banded iron formations ended and were no longer major sinks, oxygen began to replace carbon dioxide in the ocean and the atmosphere. Organisms had to develop a means of shielding their nuclei and other critical sites from oxygen or use chemical pathways that oxidized by the removal of hydrogen, rather than by the addition of oxygen. These innovations led to the evolution of the eukaryotes. With the addition of an extra step called the *citric-acid cycle* to the fermentation process (*glycolysis*), the amount of energy generated by the metabolic conversion of glucose into adenosine triphosphate (ATP) increased dramatically. ATP is the basic energy-transferring molecule with cytochrome c acting as an electron shuttle. Eukaryotes therefore only used oxygen for the production of ATP, and if other processes could yield as much ATP as does oxidation, there would be no need for oxygen at all.

A sufficient level of oxygen is needed for the process of *mitosis*, which involves the formation of two new nuclei during cell division. The synthesis of sterols, fatty acids, and collagen (the fibrous protein in muscles that gave rise to the metazoans) must be performed in the presence of oxygen. Therefore oxygen was responsible in large part for the evolution of higher forms of animals.

After about 1.5 billion years ago, the previously skimpy geologic record of preserved cellular remains became much better. It was as though the evolutionary processes were speeded up. It took another 0.75 billion years before the metazoans appeared in the fossil record, and by then the dissolved oxygen content of the sea was about seven percent of its present level.

This level of oxygen appears to have triggered an explosion of strange animals. The triggering mechanism was also aided by ecological stress, geographic isolation brought on by drifting continents, and climatic changes. Organisms no longer relied totally on surface absorption of oxygen, and gills and a circulatory system evolved when the oxygen level reached about 10 percent of its present the end of the Proterozoic (FIG. 3-14). From then on, there was an explosion of species that were the progenitors of the richness and variety of nature we enjoy today.

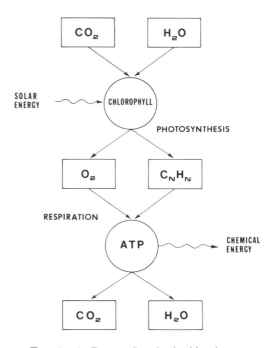

FIG. 3-13. Energy flow in the biosphere.

FIG. 3-14. Evolution of life and oxygen in the atmosphere.

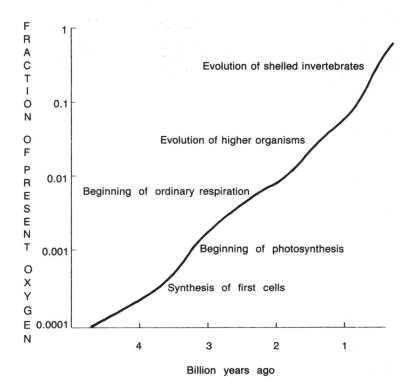

4

Age of Ancient Life

THE Paleozoic era, from about 575 to 230 million years ago, was an age of intense growth and competition in the seas and later on land, producing widely dispersed and diversified species. By the middle of the Paleozoic, all major animal and plant *phyla* (groups of organisms that share the same general body plan) were in existence. Early geologists were often puzzled why ancient rocks were devoid of fossils, and then suddenly, life appeared in rocks at the same stratigraphic horizon at the base of the Paleozoic era, in great abundance the world over.

The earliest period of the Paleozoic is the Cambrian, named for a mountain range in central Wales (FIG. 4-1), where sediments containing the earliest known fossils were found. The base of the Cambrian was once thought to be the beginning of life, and all time before was called Precambrian. The age of the Earth can be compared to the average lifetime of a man: three score and ten (70 years). The first life appeared when he was a boy of 10. By the time he was old enough to vote at age 21, the first algae turned up. When he was 30 years old, oxygen generated by photosynthesis began to accumulate in the oceans and atmosphere. At middle age, when he turned 50, the first single-celled animals came along. When he reached retirement at the age of 60, the first metazoans established themselves. In only two more birthdays, the bottom of the Paleozoic era was at hand, and thus began the ''big bang'' of life (FIG. 4-2).

PALEOZOIC ROCKS

The Paleozoic era is generally divided into two zones of nearly equal duration. The Lower (or Early) Paleozoic era consists of the Cambrian, Ordovician, and Silurian periods, (FIG. 4-3) and the Upper (or Late) Paleozoic eon consists of the Devonian, Carboniferous, and Permian periods. The first half of the Paleozoic was relatively quiet in terms of geological processes, with little mountain building, volcanic activity, glaciation, and extremes in climate. The land was divided into two supercontinents. The northern lands, called *Laurasia*, included what is now North America, Greenland, Europe, and Asia. The southern landmass, called *Gondwanaland*, in-

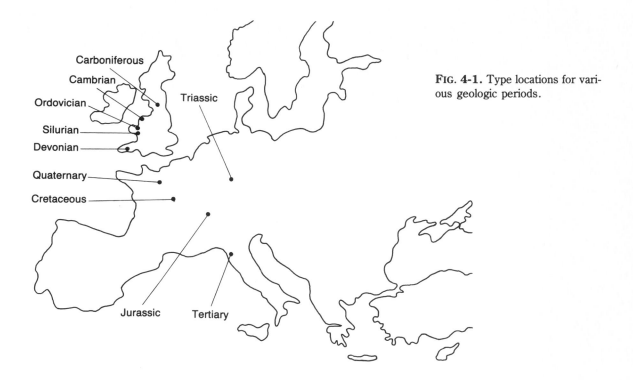

FIG. 4-1. Type locations for various geologic periods.

FIG. 4-2. Early Paleozoic fossils on display at the Museum of Geology, South Dakota School of Mines at Rapid City.

FIG. 4-3. Niagara Falls, which cuts through Silurian rocks, has been retreating at a rate of 2 to 4 feet per year.

cluded what is now Africa, South America, Australia, Antarctica, and India. The subcontinent of India later broke away from Gondwanaland and drifted into southern Asia, and the impact uplifted the Himalayan Mountains.

The two great landmasses were separated by a large body of water called the *Tethys Seaway*. Into it flowed thick deposits of sediments washed off the continents, and their weight formed a deep bulge in the ocean crust, called a *geosyncline*. The continents were lowered by erosion, and shallow seas flooded inland, covering more than half the present land area (FIG. 4-4).

Inland seas and wide continental margins, along with a stable environment, gave marine life the opportunity to flourish and become widespread. Some of the reasons why fossils were so abundant at the base of the Cambrian period are the development of hard body parts, which fossilize by replacement with either calcium carbonate or silica; long periods of deposition with little erosion, conditions that favored rapid burial (FIG. 4-5) which prevents attack

by scavengers and decay by oxidation; and large populations of species.

Just prior to the Cambrian, organisms had soft body parts, which decayed rapidly when the animal died, and only traces of their existence are found as imprints in rocks. Numerous impressions were found in the Ediacara Hills of southern Australia, which date about 670 million years in age. This was around the end of the great Precambrian ice age, the worst the world has ever experienced, when half the world was covered with ice. When the ice retreated and the seas began to warm, life took off in all different directions. There were unique and bizarre creatures, and the Cambrian saw the highest percentage of experimental organisms than any other interval in the history of Earth.

The second half of the Paleozoic followed on the heels of a Silurian ice age, when Gondwanaland wandered into the south polar region around 400 million years ago. The inland seas were crowded out and became great swamps in which coal beds accumulated during the Carboniferous period, hence its

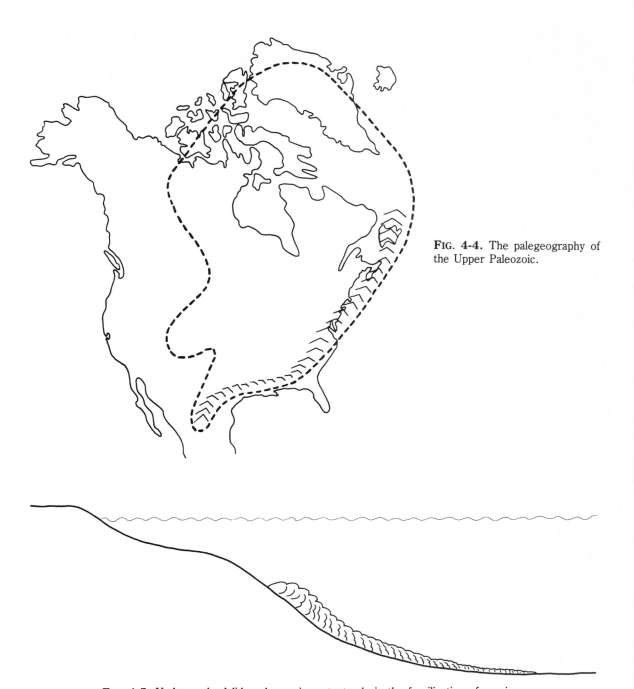

FIG. 4-4. The palegeography of the Upper Paleozoic.

FIG. 4-5. Undersea landslides play an important role in the fossilization of species.

name. The Permian period saw the complete retreat of marine waters from the land, an abundance of terrestrial red beds, and large deposits of gypsum and salt (FIG. 4-6). During the late Paleozoic era, Gondwanaland and Laurasia converged into a single crescent-shaped continent called *Pangaea* (FIG. 4-7), which extended practically from pole to pole. The sediments in the Tethys Seaway were squeezed and uplifted into mountain belts, including the Ouachitas and Appalachians of North America and the ancestral Hercynian Mountains of southern Europe. In addition, Siberia and Russia collided, creating the Urals.

The closing of the Tethys Seaway eliminated a major barrier to the migration of land plants and animals from one continent to the other. As the continents rose and the ocean basins dropped, the land became dryer and the climate became cooler. Continental margins were less extensive and narrower, placing severe restrictions on marine habitats. When the Paleozoic came to a close, the southern continents were in the grips of a major ice age, and the worst extinction event the world has ever known left the world devoid of many fascinating species.

Economically, the Paleozoic is responsible for much of the world's supply of coal (FIG.4-8). Extensive forests and swamps grew on top of and continued to add to thick deposits of peat, which were later buried under sediments. The weight of the sediments and internal heat of the Earth reduced the peat to about one-twenty-fifth its original size and transformed it into lignite, bituminous coal, and anthracite coal. Well-preserved, carbonized plant leaves are commonly found between sediment layers (FIG. 4-9).

A good portion of the world's oil reserves developed in the Paleozoic era which indicates a high degree of organic productivity in the ocean. Marine organisms buried in the shallow inland seas were covered by thick deposits of sediments. High temperatures and pressures "cooked" organic compounds into oil and gas, which migrated into subterranian reservoirs composed of porous sandstone and limestone and capped by an impenetrable rock layer.

FIG. 4-6. Major Permian evaporite deposits.

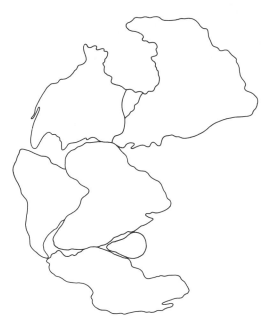

FIG. 4-7. Upper Paleozoic Pangaea.

Extensive mountain-building activity, volcanism, and granitic intrusions in the late Paleozoic provided vein deposits of metallic ores. Important reserves of phosphate used for fertilizers were laid down in the late Permian rocks in Idaho and adjacent states. Sedimentary deposits of iron were not nearly as great as those of the Precambrian period. Ore-bearing rocks of the Clinton iron formation, the chief iron producer in the Appalachian region from New York to Alabama, were precipitated by marine iron-eating bacteria.

THE INVERTEBRATES

By far the most numerous fossils representing the first abundant life on Earth were the hard parts of various marine animals that did not have backbones. The Cambrian period is best known as the "age of the trilobites" (FIG. 4-10), which appeared at the very base of the period and became the dominant species of the Early Paleozoic era. Because trilobites were so widespread and lived throughout

FIG. 4-8. Pennsylvania and Permian coal deposits in the United States.

FIG. 4-9. Fossilized plants of the Upper Pottsville series, Washington County, Arkansas.

FIG. 4-10. Cambrian-age trilobite fossils of the Carrara Formation in the southern Great Basin of California and Nevada.

the Paleozoic, their fossils became important markers for dating Paleozoic rocks. *Trilobites*, which are primitive arthropods, are ancestors to the horseshoe crab, the only remaining direct descendent alive today. The giant paradoxides, extending nearly two feet in length, were truly a paradox among trilobites.

Sharing the sea bottom with the trilobites and just as successful were the brachiopods. These animals had two saucerlike shells fitted face to face that opened and closed with simple muscles. More advanced species, including those alive today called *articulates*, had ribbed shells with interlocking teeth that opened and closed along a hinge line. The brachiopods were fixed to the ocean floor by a sort of root and fed by filtering food particles through their opened shells. The trilobites, on the other hand, had the advantage of being able to crawl around on several sets of legs in search of food, although this activity required a great deal of energy. The brachiopods conserved energy by staying in one spot, which probably contributed to their great success.

Perhaps the strangest animals ever to be preserved in the geologic record of the Paleozoic were the *echinoderms*, meaning "spiny skin." They hardly seemed like animals at all, for they had no head. The crinoids, known as sea lilies, became the dominant echinoderms of the Middle and Upper Paleozoic, and they still exist today. They had long stalks, some over ten feet in length, composed of hundreds of calcite disks. They were anchored to the seafloor with a rootlike structure. A cup called a *calyx*, which housed the digestive and reproduction systems, perched on top of the stalk. Food particles were strained from passing water currents by five feathery arms, which extended from the cup, giving the animal a flowerlike appearance.

The echinoderms are unique among the more complex animals in having five-fold radial symmetry. They are the only animals that have a system of internal canals, called a *water vascular system*, that operate a series of tube feet used in locomotion, feeding, and respiration. The great success of the echinoderms is illustrated by the fact that there are more classes of this animal living and extinct than any other phylum. The five major classes of living echinoderms include starfish, brittle stars, sea urchins, sea cucumbers, and sea lilies.

The coelenterates, which are the most primitive of animals and include sponges, jellyfish, sea anemones, sea pens, and corals, were well represented in the Lower Paleozoic. The sponges, which are the most primitive of all, were abundant, came in various shapes and sizes, and grew in thickets on the seafloor. Most coelenterates are radially symmetrical and have a saclike body and a mouth surrounded by tentacles. Most groups lacked hard parts and are rare as fossils, but the corals have hard skeletons, and successive generations have built up thick limestone reefs (FIG. 4-11).

The corals began constructing reefs in the Lower Paleozoic, forming entire chains of islands and altering the shoreline of continents. The corals built atolls atop of extinct volcanoes, and their rate of growth exactly matched the rate of subsidence of the cones beneath the sea. They captured food from the surrounding water with stingers at the end of threads, which they shot at their prey. The coral live in symbiosis with algae in their tissues helping them metabolize food. Because the algae need sunlight to grow, the coral must live close to the surface. Many corals diminished and were replaced by sponges and algae in the Late Paleozoic as a result of the retraction of the seas in which they once thrived.

Mollusks are thought to derive from the Upper Precambrian annelid worms and have lost all traces of internal segmentation. They are a diverse group and probably have left the most impressive fossil record of all. The phylum is so diverse that it is often difficult to find common features among its members. The three major groups are: snails, clams, and cephalopods. The mollusk shell is an ever-growing one-piece coiled structure for most species and a two-part shell for clams. Mollusks have a large muscular foot, for creeping, burrowing, or modified into tentacles for seizing prey.

Snails and slugs comprise the largest group and ranged throughout the Phanerozoic eon. The clams are generally burrowers, although many are also at-

FIG. 4-11. Major coral reefs.

tached to the ocean floor. The shell consists of two valves that hang down on either side of the body. Except for scallops and oysters, these valves are mirror images of each other. The cephalopods, which include the squid, cuttlefish, octopus, and nautilus, use jet propulsion to get around. They suck water into a cylindrical cavity through openings on each side of the head and expel it under high pressure through a funnel. The extinct nautiloids grew to lengths of thirty feet and more, and with their straight, streamline shells they were among the swiftest and most spectacular animals of the ancient seas.

The arthropods constitute the largest phylum of living organisms, with roughly 1 million species or about 80 percent of all known animals. They have conquered land, sea, and air and are found in every environment. The body is segmented, which suggests a relationship with the annelid worms. Paired, jointed limbs are generally present on most segments and are modified for sensing, feeding, walking, and reproducing.

The trilobites achieved the first great success at the beginning of the Paleozoic era. One giant nontrilobite arthropod found in the Middle Cambrian Burgess Shale of western Canada was a yard long.

The crustaceans are primarily aquatic and include shrimp, lobsters, crabs, and barnacles. Because crustaceans must shed their outer skeletons in order to grow, it is possible that a single individual could have left several fossils behind. The arachnids are composed of mostly air-breathing species, including spiders, scorpions, daddy long legs, ticks, and mites. One giant Paleozoic sea scorpion with massive claws grew over six feet long.

The insects are by far the largest living group of arthropods. They have three pairs of legs and typically two pairs of wings on the thorax. In order to fly, insects must be lightweight; therefore, they do not fossilize well, except when trapped in pine sap, which then fossilizes into amber.

There were a number of weird animals, some of which were possible carryovers from the Upper Precambrian and never made it beyond the Middle Paleozoic. They were so strange, they defied efforts to classify them into existing phyla. One of these animals, appropriately named *hallucigenia*, propelled itself across the seafloor on seven pairs of pointed stilts. Seven tentacles arose from the upper body, and each appeared to have its own individual mouth. Another odd animal had five eyes arranged across its head, a vertical tail fin to help steer it across the seafloor, and a grasping organ projected forward for catching prey. One worm had enormous eyes and prominent fins.

The *conodonts*, which resemble jawlike objects, are among the most puzzling of all fossils and are thought to be part of an unusual, soft-bodied animal. Graptolites were colonies of cupped organisms that resembled stems and leaves. They thus looked much like plants, but were actually animals. They either fixed themselves to the seafloor like small shrubs, floated freely near the surface, or attached themselves to seaweed. Large numbers of these organisms were buried in the bottom mud and fossilized into carbon, producing organic-rich black shales. The strata is so common the world over that graptolites are the most important group of fossils for long-distance time correlation of the Lower Paleozoic period.

A FISH STORY

The first *vertebrates*—animals with an internal skeleton—were probably wormlike creatures. Each had a prominent rod called a *notochord* along its back, and a system of nerves along the spine. Rows of muscles were attached to the backbone in a banded pattern. Rigid structures made of bone or cuticle acted as levers to translate muscle contractions, with the help of flexible joints, into organized movements (such as rapid lateral flicks of the body) to propel the animal through the water. Later, a tail and fins were added to keep the animal stable as it swam, and the body became more streamlined, like a torpedo.

With intense competition among the stationary and slow-moving invertebrates, any advancement in mobility brought with it a distinct advantage. The first protofish were jawless, generally small about the size of a minnow, and heavily armored with bony plates. Although well protected from their invertebrate enemies, the added weight kept these fish mostly on the bottom, where they sifted mud for food particles and expelled the waste through slits on both sides of the throat. These slits later became gills. Gradually, the protofish acquired jaws with teeth, the bony plates gave way to scales, lateral fins developed on both sides of the lower body, and the fish used air bladders for buoyancy, like ballast tanks on submarines. Thus, for the first time, vertebrates skillfully propelled themselves through the sea, and the fishes soon became masters of the deep.

The Devonian period is often called the "age of the fishes" (FIG. 4-12), and the fossil record reveals so many and varied kinds of fish that *paleontologists* (geologists who study fossils) have a difficult time classifying them all. Fish comprise over half the species of vertebrates, both living and extinct. They include the jawless fish (lampreys and hagfish), cartilaginous fish (sharks, skates, rays, and ratfish), and the bony fish (salmon, swordfish, pickerel, bass, and others).

All major classes of fish alive today had ances-

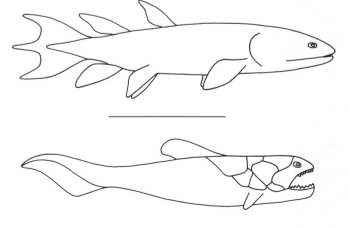

FIG. 4-12. Devonian fishes: a crossopterygian (top) and a placoderm (bottom).

tors in the Devonian, but not all Devonian fishes made it to the present. The extinct placoderms were fearsome giants, reaching thirty feet and more in length.

The *coelacanths*, called "living fossils," are an exception to the rule. Thought to be extinct for 70 million years, a six-foot coelacanth was caught in deep water off the South African coast in 1938. The fish looked ancient, a sort of castaway from the past, with a fleshy tail, a large set of forward fins behind the gills, powerful jaws, and heavily armored scales. The remarkable thing about the catch and succeeding ones was that the fish had not changed significantly in 460 million years.

The sharks (FIG. 4-13) were highly successful from the Devonian period to the present. Their ancestors replaced the bones in their skeletons with cartilage, which is a more elastic and lighter material. Even with this improvement, the shark's body is still heavier than water and the animal must keep swimming or sink to the bottom. If it needs to rest at night, it must do so on the seafloor.

When a shark charges, it cannot stop, but can only veer away. As a white shark approaches its prey, its eyes focus backwards and are covered by a membrane that shields them from the shark's flailing victim. White sharks mistakenly attack people thrashing in the water because what they initially see looks like prey from a distance. As the shark gets closer, instead of seeing its target clearly and veering away, it loses its sight and continues to charge blindly. Like all fish, sharks have a keen sense of smell and can zero in on a wounded, bleeding animal from afar. Therefore, several sharks might converge on a single kill and a furious fight for possession then ensues.

A shark breathes by taking in water through its mouth, passing it over the gills, and expelling it through distinctive slits in its sides.

Closely related to the shark is the ray, which became greatly flattened with pectoral fins enlarged into wings and a tail reduced to a thin, whiplike structure. Rays glide through the water by flapping their wings, which can be twenty feet or more across.

Connecting links between the fishes and the air-breathing vertebrates, are found in the Devonian crossopterygians, which are extinct ancestors to terrestrial vertebrates and the lungfish, another "living fossil". The crossopterygians are lobe-finned, which means that the bones in their fins are attached to the skeleton and arranged into primitive elements of a walking limb. They could breathe by taking air into primitive nostrils and lungs as well by using their gills, thereby placing them in the direct line of evolution from fish to land-living vertebrates. Air-breathing was also important to fish trying to survive in warm, shallow, oxygen-poor waters.

Modern lungfish live in African swamps that

FIG. 4-13. The shark has been in existence since the Devonian period.

seasonally dry out, forcing the fish to hole up for long stretches until the rains return. They hole up by burrowing into the moist sand, leaving an air hole leading to the surface. They breathe by pumping air into primitive lungs, and live in suspended animation. Thus, they can survive out of water for several months or even a year or more if necessary. When the rains return and the water fills the pond again, the fishes come back to life and breathe normally through their gills. In Florida, there is a species of walking catfish, members of which will leave their drying pond and walk by pushing themselves along with tail and fins sometimes a considerable distance until they find another suitable home.

CONQUERING THE LAND

Animals of the Proterozoic era were so varied and interesting that often the plants are overlooked. The distinction between plants and animals is blurred in the geologic record of the obscure past when they shared many common characteristics. From their humble Precambrian beginnings as simple algae, single-celled plants probably colonized for much the same reasons single-celled animals did: mutual support, division of labor, and protection.

Like the animals, plants did not show up in the fossil record as complex organisms until the Cambrian period, after which they began to evolve rapidly. The early seaweeds were soft and nonresistant, and for the most part did not fossilize well. Nonetheless, the Cambrian has been described as ''the age of seaweed,'' even though the geologic record does not support this contention with strong fossil evidence. A variety of fossil spores have been found from Precambrian and Cambrian sediments (FIG. 4-14), suggesting complex sea plants, but there are no other significant remains. Even Ordovician plant fossils appear to be composed almost entirely of algae, probably similar to present-day algal mats found on seashores and the bottoms of ponds.

During the Ordovician period around 450 million years ago, the concentration of oxygen in the atmosphere generated sufficient levels of ozone in the upper stratosphere to shield the Earth from the Sun's deadly ultraviolet rays, and for the first time, plants came ashore to populate the land (FIG. 4-15). Some marine algae managed to live in the intertidal zones. They could stay out of water for only a short time, however; otherwise, they would dry out and die. Even with protective measures to help the organisms survive out of water for longer periods, they were still dependent on the sea for reproduction.

(Courtesy of National Park Service)

FIG. 4-14. Cambrian-age sandstone at Pictured Rocks, Michigan.

Table 4-1. Geologic Time Scale.

Eon	Era	Period	Epoch	No. Years Ago Began (In Millions Of Years)	First Life Forms	Geology
Phanerozoic	Cenozoic	Quaternary	Holocene	.01		
			Pleistocene	2	Man	Ice age
			Pliocene	5	Mastodons	Cascades
		Tertiary	Miocene	24	Apes	Alps
			Oligocene	37	Saber-tooth tigers	
			Eocene	54	Whales Horses	Rockies
			Paleocene	65	Alligators	
	Mesozoic	Cretaceous		137	Birds	Sierra Nevadas
		Jurassic		180	Mammals	
		Triassic		230	Dinosaurs	Atlantic
	Paleozoic	Permian[1]		290	Reptiles	Pangaea
		Pennsylvanian[1,2]		320	Trees	
		Mississippian[1,2]		350	Amphibians Insects	
		Devonian		405	Sharks	
		Silurian		435	Land plants	
		Ordovician		500	Fish	
		Cambrian[3]		575	Sea plants	
Protozoic	Proterozoic			2500	Invertebrates	Ice age
	Archean[4]			4600	Prokaryotes	First rocks

[1]Time between these two periods is called Permocarboniferous time.
[2]These two periods are also together called the Carboniferous period.
[3]All time before this period is known as Precambrian time.
[4]This eon is also called the Archeozoic eon.

Lichens, which are a partnership between algae and fungus whereby one lives off the waste products of the other, probably took the first tentative steps on dry land. They were followed by the mosses and liverworts. A seaweedlike plant grew half in and half out of the water in estuaries and rivers. For plants to be truly shore-bound, they had to be able to reproduce out of water. The first land

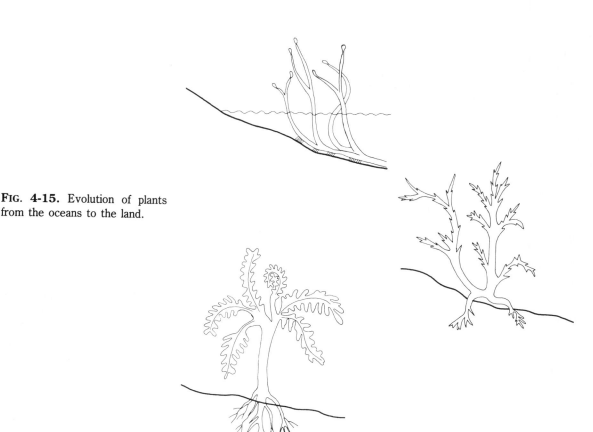

Fig. 4-15. Evolution of plants from the oceans to the land.

plants did so by producing sacs of spores attached to the ends of simple branches. When the spores matured, they were cast into the air and carried by the wind to suitable sites, where they grew into new plants.

The next major step came with the development of a vascular stem that could conduct water from a swamp or the nearby moist ground to the extremities. The structure gave the plants strength and enabled them to stand up. The early club mosses, ferns, and horsetails were the first plants to make use of this system. True roots developed and allowed plants to survive on dry land by taking advantage of the soil moisture, which was replenished by rain.

When the true leaves developed, branches had to be continuously strengthened as the leaves grew larger; otherwise they would mechanically fail as the tree grew longer and heavier. Also, in order to maximize photosynthetic activity, leaves had to be exposed to as much sunlight as possible, which placed further mechanical stress on the plant. The plants that developed a branching pattern, which gathered the most light, were the most successful; therefore, competition for light was of primary importance in the evolution of land plants. The best branching pattern was composed of tiers of branches, in which the higher branches become successively shorter with a minimum of self-shading. This pattern immerged during the first 50 million years of land plant life and continues today.

The first invertebrates to crawl out of the sea were probably *crustaceans*, which were segmented creatures, and ancestors of today's millipedes. They had a highly successful means of locomotion and walked on perhaps a hundred pairs of legs. They

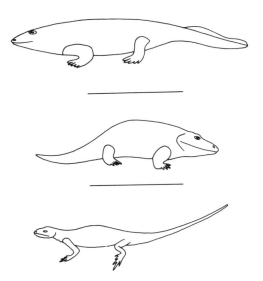

FIG. 4-16. The earliest amphibians.

stayed close to shore at first and moved farther inland along with the mosses and lichens. With no competitors, these creatures had the land to themselves, and with plenty to eat, they became the first land giants, upwards of six feet in length. This situation did not last long. When relatives of the giant sea scorpion, which terrorized the seafloor with immense pincers, took to the land, the ancestral millipedes became easy prey.

The advent of the forests, when leaves and other edible parts were no longer close to the ground within easy reach, posed new problems for the ancestors of the insects. Climbing up tall tree trunks to get at the leaves was probably easier than getting back down. It might have been better to simply jump or glide down on primitive winglike structures, which originally were intended to regulate the insect's body temperature. By natural selection, these structures developed into flapping wings. They worked well for launching insects up to the tree tops and also came in handy for escaping predators like the vertebrates that had come ashore.

The plants had been greening the Earth for almost 100 million years before the vertebrates finally set foot on dry land. By the Middle Devonian, about 370 million years ago, stiff competition in the sea encouraged crossopterygians and their kin to make short forays on shore for crustaceans and insects, which were in abundance. Digging in the sand for food and shelter strengthened their lobe fins into walking limbs, allowing them to venture farther inland, although not too far from accessible swamps or streams. By the Late Devonian, these fish gave rise to the earliest amphibians (FIG. 4-16).

The amphibians still depended on a nearby source of water to keep their skins moist and for reproduction. They were slow and ungainly, and their weak legs could hardly hold their squat bodies off the ground for any length of time. In order to feed, the early amphibians developed extendable tongues that lashed out at prey and flicked them into their mouths, thereby becoming highly successful hunters without the need for speed or agility. The necessity of living a semiaquatic lifestyle led to the eventual downfall of the amphibians when the great swamps dried up toward the end of the Paleozoic era. The void left by the amphibians was quickly filled by their second cousins, the reptiles, which were better adapted for a life totally out of water and were destined to become the greatest success story the world has ever known.

5

Age of Middle Life

THE Triassic, Jurassic, and Cretaceous periods comprise the Mesozoic era, which occurred from 230 million to 65 million years ago. The Earth had just come out of the Late Paleozoic ice age and an extinction event that took 95 percent of all animals and plants. Animals that were immobile and could not migrate to a better habitat and those that developed specialized life-styles and were unable to adapt to a changing environment were hit hardest. The animals that were able to cross the thin red line from the Paleozoic to the Mesozoic era became anatomically different from their close relatives that were left behind. Thus, the bottom of the Mesozoic was a sort of new beginning of life.

At the start of the Mesozoic, all the continents were consolidated into a supercontinent; about midway, they began to break up; and at the end, most continents were well on their way to their present positions. There were three major bodies of water—the Atlantic, Arctic, and Indian oceans—that were not in existence when the era commenced. The climate was exceptionally mild for an unusually long period of time. One group of animals that took advantage of these excellent conditions grew to enormous size. Some returned to the sea, while others took to the air. Then, for yet unexplained reasons, these giant reptiles disappeared, along with 70 percent of all other species, and life was again forced to start anew.

MESOZOIC GEOLOGY

At the beginning of the Triassic period, there was a single large continent and a single large ocean. The great glaciers of the previous ice age melted, and the seas began to warm. The high mountain ranges of North America and Europe were lowered by erosion. Reef building was intense in the Tethys Seaway, and thick deposits of limestone and dolomite laid down by lime-secreting plants and animals were later uplifted to form the Alps.

In North America, terrestrial red beds covered the Colorado Plateau and the region from Nova Scotia to South Carolina. They were also common in Europe. The wide occurrence of red sediments might have been the result of large amounts of iron supplied by intense igneous activity the world over. Air bubbles trapped in ancient tree sap also indicate that there was a greater abundance of oxygen in the atmosphere, which oxidized the iron.

In Siberia, there were great lava flows and granitic intrusions. Extensive lava flows also covered South America, Africa, and Antarctica. Southern Brazil was covered with 750,000 square miles of basalt, constituting the largest lava field in the world. These great outpourings probably reflected enormous crustal movements. By the close of the Triassic, North and South America separated; India, which was nestled between Africa and Antarctica, began to separate; and a great rift began to separate the North American continent from Eurasia, which would eventually form the North Atlantic Ocean.

During the Jurassic period, which began roughly 180 million years ago, North America continued to drift westward, while the North Atlantic continued to widen at the expense of the Pacific. South America began to separate from Africa, and India was fully adrift and headed toward southern Asia. Antarctica, still attached to Australia, swung away from Africa toward the southeast, and the proto-Indian Ocean was formed. The Tethys Seaway provided a wide gulf between the northern and southern landmasses and continued to fill with thick layers of sediment.

An interior seaway called the Sundance flowed into the west-central portions of North America, and accumulations of marine sediments that were eroded from the Cordilleran highlands to the west were deposited on the terrestrial red beds of the Colorado Plateau, forming the Morrison Formation, which is known for its abundance of dinosaur bones (FIG. 5-1). Eastern Mexico, southern Texas, and Louisiana were also flooded. In South America, great floods of basalt upwards of 2,000 or more feet thick covered large parts of Brazil and Argentina. Basalt flows also occurred in Africa and Antarctica, and from Alaska to California, along with granitic intrusions such as the huge Sierra Nevada batholith of California. Because of the effects of seafloor spreading, in which new oceanic crust was created at mid-ocean ridges and old oceanic crust was consumed

deep ocean trenches, the oldest rocks found in the Pacific Basin are Upper Jurassic in age.

In the Cretaceous period, great deposits of limestone and chalk were laid down in Europe and Asia. Seas invaded Asia, South America, Africa, Australia, and especially the interior of North America (FIG. 5-2). Into this great seaway were deposited thick layers of sediment, which are presently exposed as impressive cliffs in the western United States (FIG. 5-3).

The Appalachian Mountains which were an imposing mountain range at the beginning of the Triassic, were eroded down to stumps by the Cretaceous. The rim of the Pacific Basin became geologically highly active, and practically all the mountain ranges facing the Pacific and the island arcs along its margins developed during this period. Toward the end of the Cretaceous, North America and Europe were no longer in contact, except for a land bridge created by Greenland in the north. The strait between Alaska and Asia narrowed, creating the Arctic Ocean, which was practically landlocked. The South Atlantic continued to widen, separating South America and Africa by over 1500 miles of ocean. Africa moved northward, leaving Antarctica (still joined to Australia) behind, and began to put the squeeze on the Tethys Seaway. Meanwhile, India narrowed the gap between it and southern Asia. As Antarctica and Australia moved eastward, a rift developed that would eventually separate them.

The position of the continents might have been a major contributing factor to the warm Cretaceous climate (FIG. 5-4), in which global average temperatures were 10 to 25 degrees Fahrenheit warmer than they are today. There were no large variations in temperature between the tropics and the poles, and there is no evidence of any permanent ice sheets in the polar regions. The deep ocean water, which is now near freezing, was 60 degrees Fahrenheit warmer during the Cretaceous period. Coral reefs and other tropical biota that require warm water ranged as much as 1,000 miles closer to the poles than they do today. Alligators and crocodiles thrived in the higher latitudes as far north as present-day Labrador, whereas today, they do not range much farther north than the Gulf Coastal states. Polar

FIG. 5-1. (top) The Jurassic Morrison Formation, Uinta Mountains, Utah.
(bottom) Restoration of dinosaur bones at Dinosaur National Monument, Utah.

Mesozoic Geology 61

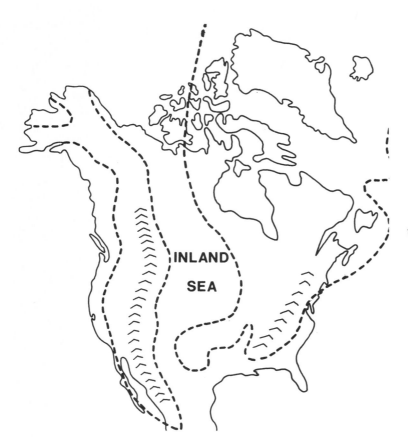

FIG. 5-2. The paleogeography of the Cretaceous period.

INLAND SEA

(Courtesy of National Park Service)

FIG. 5-3. Sandstone cliffs of Late Cretaceous age in Mesa Verde National Park.

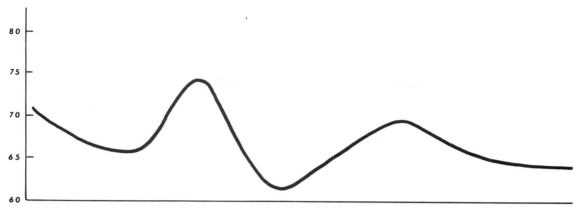

FIG. 5-4. Variation in worldwide temperatures during the Cretaceous.

forests extended into latitudes 85 degrees north and south of the equator. The most remarkable example is a well-preserved fossil forest on Alexander Island in Antarctica.

Continents congregating in the low latitudes near the equator allowed warm ocean currents to carry heat poleward. In addition, oceans in the polar regions reflected heat less than land and therefore absorbed more heat. The oceans also were interconnected in the equatorial regions by the Tethys and Central American seaways, which provided a unique circumglobal current. The continents were flatter with lower mountains and higher sea levels. Perhaps the greatest contribution to the warming of the Earth came from increased volcanic activity brought on by the most active plate tectonics the world has ever known. The volcanic activity produced large amounts of atmospheric carbon dioxide which substantially increased the greenhouse effect. It also provided an abundant source of carbon for green plants which contributed to their prodigious growth.

The mineral resources of the Mesozoic era are possibly the most important to modern industrialized nations. Over half the world's known petroleum reserves are in Mesozoic rocks. Huge quantities of oil and gas are pumped out of Mesozoic sediments underlying the Middle East, the Gulf Coast region, the Rockies, the North Slope of Alaska, and the North Sea. The great Mesozoic swamps produced thick coal deposits in the western United States (FIG. 5-5) which rival those of the Carboniferous period. Abundant deposits of coal are also found in Canada, South Africa, and Asia. Mesozoic Porphery copper deposits were formed in the Cordilleran regions of North and South America. Triassic and Jurassic rocks in the western United States contain important reserves of uranium, which were responsible for the great uranium boom from the 1950s to the 1980s. The mother lode of the California gold rush days is Late Jurassic or Early Cretaceous in age. The diamond-bearing Kimberlite pipes of South Africa are probably of Cretaceous age, although the diamonds they contain most likely formed at an earlier time.

LIFE IN THE MESOZOIC ERA

The Mesozoic era was a time of transition, especially for the plants. Plants at the beginning of the era showed little resemblance to those at the end, which were more closely related to the plants of today. The gymnosperms, which began in the Permian, bore seeds with no fruit covering. The true ferns prospered even in the temperate climates, but now they are restricted to the tropics. The cycads, which resembled palm trees, were also highly successful and ranged across all the major continents, possibly contributing to the diets of plant-eating dinosaurs. The ginkgo, of which the maidenhair tree in eastern China is the only relative alive today,

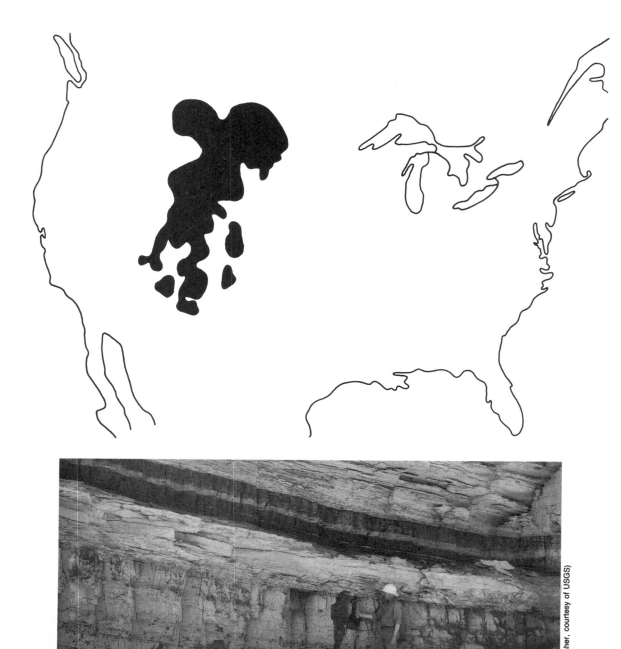

FIG. 5-5. (top) Cretaceous and Tertiary coal deposits in the United States. (bottom) Cretaceous coal seam at the Sunnyside coal mine, Carbon County, Utah.

(Photo by D.J. Fisher, courtesy of USGS)

might be the oldest living genus of seed plants. Also dominating the landscape were the conifers, whose petrified trunks (FIG. 5-6) are as much as 5 feet across and a 100 feet long.

A sudden change in vegetation came in the Middle Cretaceous period with the introduction of the angiosperms, which evolved alongside pollinating insects. They were distributed worldwide by the close of the Cretaceous, and today include some 250,000 species of trees, shrubs, grasses, and herbs. The plants offered to pollinators brightly colored and scented flowers, the promise of sweet nectar, and even flower parts that mimicked insects. The unwary intruder was dusted with pollen, which was transported to the next flower the insect visited, thus pollinating it. Many angiosperms also depended on animals to spread their seeds, which were encased in delicacies that were eaten by the animal, passed through the body, and dropped some distance away. The rise of the angiosperms might even have contributed to the extinction of the dinosaurs and certain marine species at the end of the Cretaceous because the angiosperms absorbed large quantities of carbon dioxide from the atmosphere, resulting in a drop in global temperatures.

Early in the Triassic period, the ocean was still cool from the latest ice age, and those marine invertebrates that managed to escape extinction lived in a narrow margin near the equator. Corals, which require warm, shallow water, were particularly hard hit as evidenced by the fact that no coral reefs of Early Triassic origin have been found. The crinoids and brachiopods, which had their heyday in the Paleozoic, were relegated to minor roles. The trilobites were even less lucky and became extinct at the close of the Permian.

The space vacated by the trilobites was taken over by a variety of other crustaceans, including shrimps, crabs (FIG. 5-7), crayfish, and lobsters. The mollusks appeared to have weathered the hard times of the Late Paleozoic quite well and went on to become the most important shelled invertebrates of the Mesozoic seas. Some 60,000 species are alive today. There were giant clams up to three feet long, giant squids up to 65 feet long and weighing over a ton, and crinoids 60 feet long. The cephalopods

(Photo by Richard Frear, courtesy of National Park Service)

FIG. 5-6. Petrified Forest, Arizona.

became the most spectacular, diversified, and successful marine invertebrates of the Mesozoic.

The ammonites (FIG. 5-8), which are identified by their complex suture patterns and were as much as 7 feet across, are the most important guide fossils for dating Mesozoic rocks. Unfortunately, after making it through the critical transition from Permian to Triassic and recovering from serious setbacks in the Mesozoic, the ammonites suffered final extinction at the close of the Cretaceous.

Among the vertebrates, fish progressed from having rough scales, asymmetrical tails, and cartilage in their skeletons to having flexible scales, powerful advanced fins and tails, and all-bone skeletons much like the fish of today. The sharks regained ground lost during the great extinction and went on to become the successful predators they are today. The dolphinlike ichthyosaurs, the sea-cowlike placodonts, and the sea-serpentlike plesiosaurs were actually reptiles that returned to the sea, where they achieved great success. The lizards and turtles also went to sea in the Cretaceous. Of all the marine rep-

tiles, only the smallest turtles made it past the extinction at the end of the period.

The amphibians continued to decline in the Mesozoic with all large, flat-headed species becoming extinct early in the Triassic. The group thereafter was represented by the more familiar toads, frogs, and salamanders. Although the amphibians did not achieve complete dominion over the land, their descendants, the reptiles, became the rulers of the world, and the Mesozoic is more popularly known as the "age of reptiles."

RISE OF THE REPTILES

In the latter part of the Paleozoic era, the reptiles largely replaced the amphibians, and they became the dominant land-dwelling animals of the Mesozoic. One of the reptiles' claim to fame was their efficient means of locomotion. The improvement over the amphibian foot included changes in the form of the digits, the addition of a short, thumblike fifth digit, and the appearance of claws.

FIG. 5-7. Mesozoic fossils on display at
the Museum of Geology, South Dakota School of Mines at Rapid City.

In some reptiles, the stance narrowed and the stride lengthened. Others maintained a quadrupedal gait when moving slowly, but reared up on their hind legs when running. Their body pivoted at the hips and their long tail counterbalanced their nearly erect trunk. This stance probably freed their forelimbs for attacking small prey and other useful purposes. Reptiles have scales, which retain the animal's bodily fluids, whereas amphibians have permeable skin, which absorb water and oxygen. Amphibians will dry out if away from water for too long.

Another of the reptile's improvements over the amphibians were its eggs. Amphibians, like fish, lay their eggs (which do not have waterproof shells) in water or at least in moist places, and the young must fend for themselves when they hatch. Reptiles, on the other hand, lay their eggs (which have hard, watertight shells) on dry land and tend to their young, which gives them a better chance of survival and which contributed to the reptiles great success in populating the land.

Like fish and amphibians, reptiles are cold blooded, which is really a misnomer since they draw heat from their environment. The blood of a reptile sunning on a rock can be warmer than that of a mammal. One ancient reptile called *dimetrodon* even had a huge sail on its back, the purpose of which was apparently to regulate its body temperature by absorbing sunlight when it was cold and radiating excess body heat when it was hot. Having a high body temperature is as important to reptiles as it is to mammals, and in the cold mornings the animals are sluggish and must wait until the Sun warms their bodies before their metabolism can operate at peak performance. For this reason, reptiles only need about one-tenth the amount of food that mammals need to survive; mammals require most of their calories just to maintain their body temperatures. As a consequence, reptiles can live in deserts and other desolate places and flourish on small quantities of vegetation that could not keep a mammal the same size alive for very long. The generally warm

FIG. 5-8. Cretaceous ammonite fossils on display at the Museum of Geology, South Dakota School of Mines at Rapid City.

climate of the Mesozoic era was very advantageous to the reptiles and greatly aided them in colonizing the land, whereas the amphibians, which had to avoid direct sunlight, were relatively cold and slow-moving.

At the close of the Triassic period, reptiles were the leading forms of animal life on Earth, occupying land, sea, and air. During this time, a remarkable reptilian group appeared in the fossil record. They were the alligators with a blunt head, the crocodiles with a more elongated head, and the gavials with an extremely narrow head. Members of this group adapted to life on dry land, a semiaquatic life, or an entirely aquatic life with a sharklike tail, a stream-lined head, and legs remolded into swimming pad-dles, and the animals were about fifteen feet long.

The crocodilians have diversified considerably since the Early Triassic, spreading to all parts of the world and adapting to a wide variety of ecological niches. They belong to the same subclass, Ar-chosauria (which literally means "ruling reptiles"), as the dinosaurs and flying pterosaurs. They are the only surviving members following the "great dying" at the end of the Cretaceous. Fossil crocodiles were among the first vertebrates unearthed by paleotol-ogists of the early nineteenth century, and were used as evidence for Darwin's theory of evolution. One fossil of a gaviallike monster from the Lower Cretaceous in Niger was about 35 feet long. Today, crocodiles are the largest of all reptiles and are among the only living relatives of the dinosaurs.

THE TERRIBLE LIZARDS

There are a number of reasons why animals ob-tain a large size other than to have dominance over other animals. Large reptiles possess the power of almost unlimited growth and never cease growing entirely, but continue to grow slowly until disease or accident takes their lives. The giant komodo (dragon) lizards of southeast Asia, for example, grow to over 300 pounds and prey on monkeys, pigs, and deer.

With their continued growth, reptiles achieve a measure of eternal youth, whereas mammals grow rapidly to adult age and then slowly degenerate and die. For reptiles, a large body helps them retain their heat for long periods. A large body retards heat loss better than a small one because it has a smaller sur-face area per volume. As a result, the animal is less susceptible to short-term temperature variations in its environment, such as during a cool night or a cloudy day. Conversely, it also takes much longer for a large reptile to warm up from an extended cold period than it does for a small one. A certain amount of body heat is also generated in the muscles, al-though this is only about a quarter of what mammals can generate. A high, steady body temperature keeps the metabolism in top running condition at all times, and the output of muscles is greater at high temperatures than at low temperatures; therefore, the performance of dinosaurs probably could match that of large mammals.

The generally warm climate of the Mesozoic era produced lush vegetation, including ferns and cycads, that must have been very appealing to the plant-eating dinosaurs. These dinosaurs had to develop large stomachs in order to digest the tough, fibrous fronds. Some species swallowed rounded cobbles called *gizzard stones* to pulp the coarse vegetation in their stomachs. The digestive juices further broke the material down, and the long process of fermen-tation required a huge storage vat. A large stom-ach, therefore, requires a large body to carry it around, as seen in today's large ungulates, the rhi-noceros and elephant. Carnivorous dinosaurs needed to grow to considerable stature so they could prey on the giant herbivores.

About the only thing that kept the dinosaurs from growing larger than they did was the force of gravity. As these animals doubled in size, the weight the bones needed to support quadrupled. The ex-ceptions were those dinosaurs that took to the sea and present-day whales. The buoyancy of the wa-ter kept the weight off their bones.

Many of the smaller, earlier dinosaurs reared up on their hind legs and became the first animals to establish a successful, permanent, two-legged stance (FIG. 5-9). Walking on two legs increased their speed and agility and freed their forelimbs for foraging and other tasks clumsy legs were unable to do. However, it also meant that the back legs and hip had to support the entire weight of the animal, while a large tail counterbalanced the upper portions of the body. These animals walked birdlike and, thus, all dinosaurs are divided into two major groups (FIG. 5-10): ornithischians, with a birdlike pelvis, and saurischians, with a reptilelike pelvis.

Some of the bipedaled dinosaurs later reverted to a four-footed stance, probably because of increased weight, eventually giving rise to the gigantic, long-tailed, long-necked sauropods such as the apatosaurus (formerly called *brontosaurus*). Others, like tyrannosaurus rex, the greatest land carnivore of all times, kept their two-legged stance and stood on powerful hind legs, with arms shortened to almost useless appendages. The ornithischians probably arose from the same group of thecodont reptiles that gave rise to birds and crocodiles. Indeed, birds have been called the only living direct descendants of the dinosaurs, and the skeletons of the smaller, lighter dinosaurs bear many resemblances to those of birds.

There is growing evidence that dinosaurs might have been warm-blooded. The fact that dinosaurs and birds, which are warm-blooded, had a common ancestor, the thecodonts, makes a strong argument for this theory. Once described as large, lumbering animals because of the manner in which their bones were reconstructed at museums, dinosaurs now appear to have been highly predaceous, extremely agile, and very active animals. These abilities require a high metabolic rate which in turn means they needed to be warm-blooded.

At the beginning of the dinosaur age, the climates of southern Africa and southern South America, where the dinosaurs once roamed, had cold winters that large, cold-blooded animals would not be able to survive, unless they migrated yearly to warmer climates. The stamina needed for such a migration would require sustained energy levels that only warm-bloodedness could supply. Warm-blooded animals grow to adulthood more rapidly than cold-blooded animals, and a comparison among the bones of dinosaurs, birds, and crocodiles—all of which had a common ancestor—show there is a remarkable similarity between bird and dinosaur bones. In addition, the dinosaur bones had an even

FIG. 5-9. Fossil dinosaur footprint on Dakota sandstone, Jefferson County, Colorado.

(Photo by J.R. Stacy, courtesy of USGS)

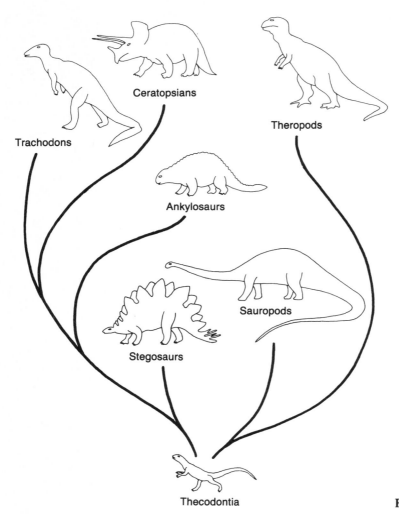

Trachodons

Ceratopsians

Theropods

Ankylosaurs

Sauropods

Stegosaurs

Thecodontia

FIG. 5-10. Family tree of dinosaurs.

vessel density than those of currently living mammals. Some dinosaur skulls show signs of sinus membranes, which are found only in warm-blooded animals.

Perhaps no reptile has captured the imagination more than the pterosaur (FIG. 5-11), and with wingspans up to forty feet and more, they are thought to be the largest animals ever to fly. To achieve this ability, the fourth finger of each forelimb was greatly elongated and supported the front edge of a membrane that stretched from the flank of the body to the tip of the finger. Pterosaurs resembled both

birds and bats in their overall structure and proportions, and like birds they had hollow bones to conserve weight. The larger pterosaurs were much the same proportions as a modern hang glider and weighed about as much as the pilot would. They might have attained flight by jumping off a cliff and riding the updrafts, by falling from a tree in a gentle breeze, or by gliding across the tops of wave crests like modern-day albatrosses. The animal might also have trotted along the ground flapping its wings and taking off gooney-bird fashion, or it might have simply stood on its hind legs, caught a strong breeze, and with a single flap of its huge wings and a kick of its legs become airborne. Landing was probably much easier, and like a hang glider, it simply stalled near the ground and touched down gently on its hind legs.

Why pterosaurs took to the air in the first place is still a mystery. Their ancestors might have grown skin flaps for jumping out of trees like flying squirrels, or the wing membranes originally could have served to cool the animal, like a flapping fan. In either case, that the animals could fly is in little doubt, and they went on to become the greatest aviators the world has ever known.

Birds descended from the thecodonts, which also gave rise to the dinosaurs and crocodiles. As a result, birds are called "glorified reptiles." They are warm blooded in order to obtain maximum metabolic efficiency needed for flight, but they retain the reptilian mode of reproduction by laying eggs.

Birds first appeared in the Jurassic period about 150 million years ago. Archaeopteryx (Fig. 5-12), about the size of a crow, is the earliest known fossil bird and appears to be a species in transition between reptiles and birds, for although it definitely had feathers, it could not fly in the normal sense and probably just glided for short distances. The animal had teeth, claws, a long tail, and many of the skeletal features of a small dinosaur, which also included the lack of hollow bones. Its feathers were outgrowths of scales and used originally for insulation.

After mastering the skill of flight, birds quickly radiated into all environments, and their better adaptability allowed them to compete successfully with the pterosaurs, which probably contributed to the pterosaurs extinction. Giant flightless land birds appeared very early in the history of birds, and the wide distribution of their grounded descendents is further proof for the existence of a single supercontinent since the animals would have had to walk from one part of the world to another, whereas flying birds can cross entire oceans easily.

The success of the dinosaurs is exemplified by the extensive range of ecological niches they have occupied, in which they have dominated all other forms of land-dwelling animals. About 500 species

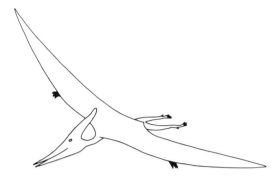

FIG. 5-11. A Jurassic pterosaur.

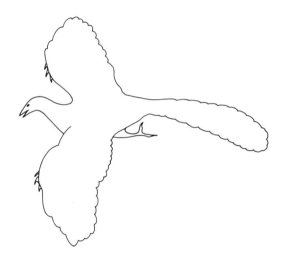

FIG. 5-12. A Jurassic archaeopteryx.

have been described, although this is probably only a fraction of the total.

The oldest dinosaurs stemmed from Gondwanaland after the last of the glaciers had disappeared. They are known to have ventured to all major continents, and their distribution is taken as evidence for continental drift (Fig. 5-13). When the dinosaurs originated during the Triassic, all the continents were still assembled into the huge landmass of Pangaea. During the Jurassic, the continents split apart. Except for a few temporary land bridges, the oceans provided an effective barrier to any further dinosaur migration. At this time, almost identical

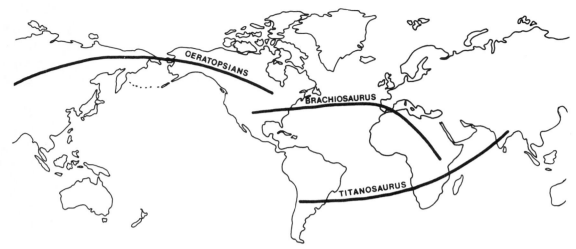

FIG. 5-13. Distribution of dinosaurs as evidence for continental drift.

species lived in North America, Europe, and Africa. The greatest dinosaur that ever existed, brachiosaurus, is found only in Colorado-Utah, southwestern Europe, and East Africa, and must have traveled to Africa by way of Europe. The splitting up of Pangaea also might have contributed to the demise of the dinosaurs by changing global climate patterns and bringing on unstable weather conditions.

Dinosaurs generally are portrayed as being unintelligent, slow-moving brutes. However, it appears that the giant herbivores traveled in herds with the largest adults in the lead and the juveniles kept in the middle for protection. Females of some species might even have given live birth to their young. Many nurtured and fiercely protected their young until they could fend for themselves. The parents might have brought food to their young and regurgitated seeds and berries like some birds do today. This parental care for the young indicates strong social bonds and might have been a major reason dinosaurs were so successful.

Some dinosaurs might have developed complex mating rituals, and it has even been suggested that the large sail on dimetrodon's back was used to attract females. Other dinosaurs might have had elaborate head gear for much the same reason.

The carnivores were cunning, aggressive creatures that charged at their victims with great agility and speed. As for brains, the relatively large cranial capacity of some carnivores suggests they were fairly intelligent and able to react to environmental pressures, which helps explain why dinosaurs were able to dominate the planet for 140 million years.

EARLY MAMMALS

The first animals to depart from basic reptilian stock, about 300 million years ago, were the pelycosaurs, which were distinguished by their large size and varied diet. Some, like dimetrodon (FIG. 5-14), which obtained a length of about eleven feet, had large dorsal sails composed of webs of membrane stretched across bony, protruding spines. The sails were well supplied with blood, and were probably used for heating and cooling the animal's body.

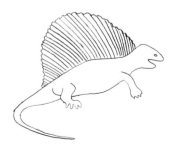

FIG. 5-14. Dimetrodon.

When the animal was cold, it turned its body broadside to the Sun so its sail could absorb more sunlight. When it was too hot, it simply faced the Sun to offer less surface area or took advantage of a gentle breeze.

Later, as the climate warmed, the pelycosaurs lost their sails and perhaps gained some degree of internal thermal control. The pelycosaurs thrived for about 50 million years and then gave way to their descendents, the mammallike reptiles called *therapsids*, which were the progenitors of mammals.

The therapsids ranged in size from as small as a mouse to as large as a hippopotamus. The early members invaded the southern continents during the Late Permian when those lands were still quite cold. This immigration is an indication that they might have been warm-blooded. They had to have some physiological adaptations to enable them to feed and move through the snows of the cold winters. They were apparently too large to hibernate, as indicated by their lack of growth rings in their bones (similar to tree rings), which indicate alternating seasons of growth. The development of fur appeared in the more advanced therapsids as they moved into colder climates. Therapsids might have also operated at lower body temperatures than most living mammals do, and thus, they were able to conserve energy by lowering their thermostat setting.

Therapsids are believed to have reproduced like reptiles by laying eggs. They might have protected and incubated their eggs and fed their young. This practice might have paved the way for longer egg retention in the mother and live births.

The therapsids dominated animal life on Earth for more than 40 million years until the middle of the Triassic and then for unknown reasons lost out to the dinosaurs. From then on, they were relegated to the role of a shrewlike, nocturnal hunter of insects until the dinosaurs became extinct.

The advantages of being warm-blooded are tremendous, and a stable body temperature finely tuned to operate within a narrow thermal range provides a high rate of metabolism independent of the outside temperature. Therefore the work output of muscles, heart, and lungs is greater, giving mammals the ability to outperform and outendure reptiles. However, the adaptation comes with a heavy price, and the animal must consume large quantities of high-caloric foods just to maintain its body temperature, forcing most species to spend the better part of their time foraging for food. The total energy requirement of birds and mammals is from 10 to 30 times that of reptiles of the same weight.

The principle of heat loss for large reptiles also applies to mammals, and a lion consumes 10 times its own weight in meat per year, whereas a shrew needs 100 times its weight in food. In addition, mammals have a coat of insulation, such as an outer layer of fat and fur, to check the escape of heat from their bodies. These animals must also have a means of getting rid of excess body heat developed during intense physical activity or on a hot day. On cold days, additional heat can be generated by *shivering*, which is spasmodic muscle contractions, or by developing other special forms of *thermogenesis*, or body heat.

Other distinguishable features of mammals include four-chambered hearts, a single bone in the lower jaw, three small ear bones, and highly differentiated teeth. They give live births and possess mammary glands that provide a rich milk to feed the young, which generally come into the world helpless. Infant mammals require a long childhood in order to grow and develop, and to learn some of the tricks of the trade from their parents. Mammals have the largest brains, capable of storing and retaining impressions. Therefore, they lived by their wits which is the reason for their great success. They conquered land, sea, and air, and are established, if only seasonally, on every corner of the globe. When the dinosaurs left the stage at the end of the Cretaceous period, the mammals were waiting in the wings ready to take over the world.

6

Age of Recent Life

BY the time the Earth entered the Cenozoic era 65 million years ago to the present, the mammals were poised to inhabit all the ecological niches left behind by the dinosaurs. It was only after the dinosaurs became extinct that the tiny creatures evolved into larger animals and became true modern mammals. Nor did the mammals cause the extinction of the dinosaurs in order to take over; they were just caught up in the natural progression from simple to complex life forms.

There are today many more numbers of diverse species than in any other period of Earth's history. Part of the reason for this great diversity is that the Cenozoic has been a time of constant change, and species have needed to adapt to a wide range of living conditions. Changing climate patterns were brought on by the movement of continents toward their present positions and intense mountain building, which raised most of the major ranges of the world.

It is not known for certain how many species occupy the Earth today, but estimates range from about 5 million to as many as 30 million. (Only a small percentage of the species which currently exist have been classified.) Nature has indeed provided us with a rich biosphere that is truly unique and most interesting.

CENOZOIC TECTONICS

The Cenozoic era is subdivided into two periods: the Tertiary which occupied most of the era, and the Quaternary, which covers the past 2 million years. Both terms are carryovers from the old geologic time scale in which the Primary and Secondary periods represented earlier Earth history. The unequal subdivision is in recognition of the unique worldwide Late Cenozoic ice age, which began about 2 million years ago.

Most European and some American geologists prefer to subdivide the Cenozoic into two nearly equal periods: the Paleogene, which spans from 65 to 26 million years ago and includes the Paleocene, Eocene, and Oligocene (FIG. 6-1) epochs; and the Neogene, which spans from 26 million years ago to the present and includes the Miocene, Pliocene.

Fig. 6-1. (top) Rugged outcrops of river-laid deposits of Oligocene age, Badlands National Park, Interior, SD. (bottom) View of Bryce Canyon National Park from Inspiration Point.

Pleistocene, and Holocene epochs. No matter which classification scheme is eventually generally accepted, the Cenozoic is regarded most everywhere as the "age of mammals."

Early in the Cenozoic, about 60 million years ago, a great rift developed between Greenland and Norway. At about the same time, Greenland began to separate from North America. Except for a few land bridges that were exposed from time to time, animals were restricted from migrating between continents. At times, Alaska connected with east Siberia at the Bering Strait (Fig. 6-2), closing off the Arc-

FIG. 6-2. The Bering Strait between Alaska and Asia.

tic Basin from warm currents and resulting in the accumulation of pack ice. The Mid-Atlantic Ridge, of which Iceland is a surface expression, began to occupy its present position midway between North America and Eurasia during the Miocene epoch, about 16 million years ago. North and South America remained separated until the Panama isthmus was uplifted during the Pliocene, about 4 million years ago, and a lively exchange of land mammals commenced. Until then, powerful currents flowed from the Atlantic into the Pacific, carrying animals from the West Indies to the Galapagos Islands, 300 miles west of Ecuador. South America was temporarily connected to Antarctica by a narrow, curved land bridge, which assisted in the migration of marsupials to Australia, which also was connected to Antarctica. Antarctica separated from Australia in the Eocene, about 40 million years ago, drifted over the South Pole, and acquired a permanent ice sheet. Most of its life forms were thus obliterated.

The Cenozoic is generally known for the intense mountain building that occurred then. The Cenozoic brought paramount changes all along the northwestern Pacific coast of North America. Volcanic activity was extensive, and great outpourings of lava

covered Washington, Oregon, and Idaho. Miocene-age basalts of the Columbia River Plateau cover 200,000 square miles and reach 10,000 feet thick. In the Late Cenozoic, the tall volcanoes of the Cascade Range (FIG. 6-3) from northern California to Canada developed, and there was extensive volcanism in the Colorado Plateau and Sierra Madre regions. The Laramide orogeny, which began about 80 million years ago and ended in the Eocene epoch, some 40 million years ago, pushed up the Rocky Mountains (FIG. 6-4), which extend from Mexico to Canada. During the Miocene, a large part of western North America was uplifted, and the entire Rocky Mountain region was raised over 3000 feet.

In the Oligocene epoch, large numbers of parallel faults sliced through the Basin and Range province (FIG. 6-5) between the Sierra Nevadas (FIG. 6-6) and the Wasatch Mountains, producing peculiar north-south trending mountain ranges. Also in the Oligocene, about 25 million years ago, a change in relative motions between the North American plate and the Pacific plate created the San Andreas Fault, which runs through southern California (FIG. 6-7). As another result of crustal movements, Baha California split off from North America and opened up the Gulf of California, and Arabia split off from Africa by a rift that formed the Red Sea.

About 50 million years ago, the Tethys Seaway narrowed as Africa approached Eurasia. The seaway began to close off during the Miocene some 20 million years ago (FIG. 6-8). Like a tablecloth thrown across a polished table, thick sediments that had been accumulating for tens of millions of years folded over into pleats, and great belts of mountain ranges formed on the northern and southern continental landmasses. More than sediments were involved; the entire crusts of both continental plates, which were up to ten miles thick, were buckled upward, forming the central parts of the range. This episode of mountain building, called the *Alpine orogeny*, ended about 30 million years ago and divided the Cenozoic into the Paleogene and Neogene periods. It raised the Alps of northern Italy, the Pyrenees between the border of Spain and France, the Atlas Mountains of northwest Africa, and the Carpathians in eastern Europe. India, which broke away from

(Photo by M.V. Shulter, courtesy of USGS)

FIG. 6-3. Mount Hood volcano of the Cascades, Hood River County, Oregon.

(Photo by George A. Grant, courtesy of National Park Service)

FIG. 6-4. The Grand Tetons.

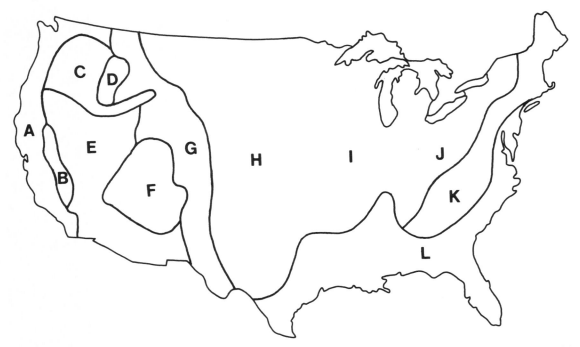

FIG. 6-5. Major geological provinces of the United States. A - Pacific Coast, B - Sierra Nevada, C - Columbia Plateau, D - Idaho Batholith, E - Basin and Range, F - Colorado Plateau, G - Rocky Mountains, H - Great Plains, I - Central Lowlands, J - Appalachian Plateau, K - Appalachian Mountains, L - Coastal Plains.

FIG. 6-6. The Sierra Nevadas, Inyo County, California.

Africa in the Early Cretaceous, sped across the ancestral Indian Ocean, slammed into southern Asia, and uplifted the Himalayans. In South America, the mountainous spine that runs along the western edge of the continent forming the Andes continued to rise throughout much of the Cenozoic as a result of the subduction of the Nazca plate beneath the South American plate.

Mineral deposits of the Cenozoic include oil, natural gas, and coal in widely scattered parts of the world. Rich deposits of metallic minerals from this period are found mostly in the Western Hemisphere. About half the world's oil fields are in Tertiary rocks, which contain nearly 40 percent of the world's oil reserves. Huge untapped reserves of oil are in the oil-shale deposits of western United States, which

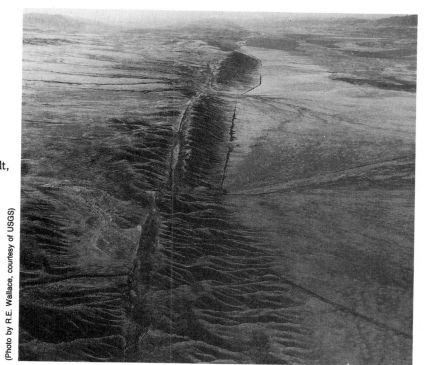

FIG. 6-7. The San Andreas fault, California.

(Photo by R.E. Wallace, courtesy of USGS)

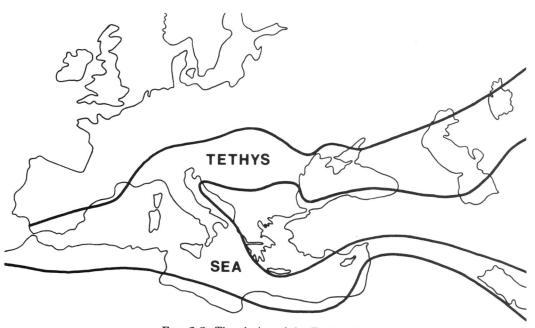

TETHYS

SEA

FIG. 6-8. The closing of the Tethys Sea.

FIG. 6-9. Tertiary rocks in the United States.

have a potential oil content exceeding that of all other resources in the world. Large reserves of coal and lignite occur in Montana, Wyoming, and the Dakotas. Most of the metal-bearing ores of North and South America are thought to be products of Tertiary igneous activity. Important deposits of copper, lead, zinc, gold, and silver are found in the Rocky and Andes mountain regions. A variety of other metallic deposits formed in the mountains of southern Europe and southern Asia during the Alpine orogeny as well. Almost everywhere, nonmetallic minerals such as sand and gravel, clay, salt, limestone, gypsum, and phosphates are mined in great quantities from Tertiary rocks (FIG. 6-9).

CENOZOIC LIFE

Extremes in climate and topography during the Cenozoic have produced a greater variety of living conditions than any other equivalent span of Earth history. The rigorous environments presented many challenging opportunities, and the extent to which plants and animals invaded diverse habitats was remarkable. The invasion of new habitats and the repopulation of the world following a major extinction are the major sources of evolutionary opportunities. As a response to those opportunities, species

show a burst of evolutionary development that gives rise to all sorts of anatomical shapes and sizes.

Nor did mammals evolve gradually from the beginning of the Cenozoic to the present, but rather in fits and starts. Most of the Paleocene was characterized by an evolutionary lag as though the world had not yet awakened from the last extinction. Then toward the end of the epoch, mammals finally began to diversify, with some large, peculiar looking animals becoming evolutionary dead ends. The mammals were cut down again during a sharp extinction event at the Eocene-Oligocene boundary about 37 million years ago, in which many of the archaic mammals abruptly disappeared.

Of the dozen or so orders of mammals that existed in the Early Cenozoic, only half were found in the proceeding Cretaceous and only half are alive today. It was only in the Eocene that most of the truly modern mammals began to appear.

Marine species that made it through the extinction event at the end of the Cretaceous looked much the same as they did in the Mesozoic. The ocean has a more moderating effect on evolution than the land because it tends to have a longer memory of environmental conditions, taking much longer to heat up or cool down. The ocean provides a better protective shield against cosmic rays and ultraviolet radiation than the atmosphere, and therefore the rate of genetic mutations is slower. Also, higher life forms like mammals evolve at higher rates than those further down the evolutionary ladder, especially those at sea.

Although the extinction in the oceans was severe and many species died out, very little in the way of radical species appeared as a result because ecological niches that were left vacant were simply taken over by the next of kin. (The situation was just the opposite on land because the dinosaurs represented the largest group of animals, and when they went, the world was left wide open to invasion by entirely new species.) Major marine groups that disappeared included the ammonoids; the rudists, which were huge coral-shaped clams; and other types of clams and oysters. All the shelled cephalopods were absent in the Cenozoic seas except the nautilus and those without shells, including cuttlefish, octopus, and squids. The squids competed directly with fish, which themselves were little affected by the extinction. Gastropods increased in numbers and variety throughout the Cenozoic, and at present, they are second only to insects in diversity.

All major groups of modern plants were

(Courtesy of National Park Service)

FIG. **6-10.** Petrified stumps on North Scarp of Specimen Ridge, Yellowstone National Park, Wyoming.

represented in the Lower Cenozoic. The *angiosperms*, or flowering plants, dominated the plant world, and all modern families appeared to have evolved by the Miocene epoch. Insects, which angiosperms are almost entirely dependent on for fertilization, continued to be of vital importance—a factor that is often overlooked. Grasses are the most important angiosperms, and there is hardly any area in which they will not grow. They provided food for mammals throughout the Cenozoic, and the grazing habits of many large mammals probably evolved in response to the availability of grass. Grain and rice are also an important staple for most of the world's human population today.

Forests of giant hardwood trees grew as far north as Yellowstone National Park, (FIG. 6-10), where now only scraggly conifers grow, indicating a present cooler climate. The cone-bearing plants,

which were prominent during the Mesozoic, occupied a secondary role during the Cenozoic. Tropical vegetation, which was widespread during the Mesozoic, withdrew to narrow regions around the equator in response to a colder, drier climate that resulted from the general uplift of the landmasses and the draining of interior seas (FIG. 6-11). Grasslands expanded as the forests retreated, and the environment grew favorable for the development of grazing animals, as well as the carnivores that preyed on them.

The drifting of the continents isolated many groups of mammals, and they evolved along independent lines. Australia is inhabited by strange egg-laying mammals called *monotremes*, including the spiny anteater and platypus, which should rightfully be classified as surviving mammallike reptiles. When the platypus was first discovered, it caused quite a

FIG. **6-11.** Paleogeography of the Upper Tertiary.

sensation in Europe. The animal had thick fur, a duck's bill, webbed feet, and a broad flattened tail. It laid eggs and was thought to be the missing link between mammals and their forbears.

Marsupials are pouched mammals whose young are so tiny after birth they have to be suckled in warm pockets on the mother's belly. They probably originated in North America around 100 million years ago, migrated to South America, crossed over to Antarctica when it was still connected to South America about 40 million years ago, and landed in Australia before it finally broke away from Antarctica. The Australian group is comprised of kangaroos, wombats, and bandicoots, while opossums and related animals occupy other parts of the world.

Camels, which originated in the early Miocene, migrated out of North America to other parts of the world by connecting land bridges (FIG. 6-12). Madagascar, which broke away from Africa in the Mesozoic, has none of the large mammals that occupy the mainland except for the hippopotamus, which mysteriously managed to cross over.

The long neck of a giraffe is often cited as a classic example of evolutionary adaptation by which a browsing mammal responded to an ever-heightening food supply. The trunk of an elephant is another adaptation for browsing. Instead of the neck lengthening, the snout was elongated to get at tall branches. The roughage the elephant ate was so poor in nutritive value that it not only needed large quantities of food, but a long time to digest it, requiring a large stomach and, like the dinosaurs, a large body to go with it. There are two living species of elephants, one of which is a close relative of the extinct mammoth. Mastodons were quite different animals (FIG. 6-13), diverging from the elephant family about 30,000 years ago. They had shorter legs and a longer body, and were covered with fine, short fur.

Another example is the horse, which seems to show a clear picture of evolutionary change through time. The earliest horses originated in western North America during the Eocene epoch and were about the size of a small dog. As time went on,

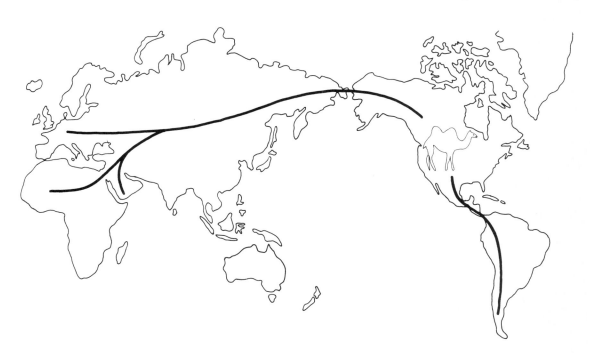

FIG. 6-12. Dispersion of the camel from North America.

FIG. 6-13. Comparison between a mammoth (top) and a mastodon (bottom).

horses became progressively larger; their faces and teeth grew longer as the animal switched from browsing to grazing; and a single toe on each foot grew into a hoof.

Bats, which have been around since the Eocene, are the only mammals that have achieved powered flight: that is, the ability to stay aloft by flapping their wings. Their wings are similar to those of the pterosaurs, except all the fingers but one are stretched out fanlike and covered with a membrane. Bats also have a unique way of finding their way in the dark: by using sophisticated sonar. Unfortunately, the adaptations needed for such elaborate equipment have made bats somewhat grotesque, which is probably why the generally harmless and beneficial creatures are unfairly disliked.

THE BRAINY BUNCH

Perhaps the greatest survival characteristic of the mammals was their relatively large brains. During the Cenozoic era, mammalian brains gradually grew larger in proportion to their bodies. As individual species grew larger in size, their brains also expanded, although not nearly at the same rate as their bodies. In addition, the size and importance of the *cerebrum*, which is the expanded anterior portion of the brain and the seat of intelligence, increased dramatically with the mammals. This increase was a result of the expansion of the *cortex*, or gray matter, which includes the outer layer of the cerebrum and the cerebellum. The cerebellum is situated between the brain stem and the back of the cerebrum and controls motor functions and body maintenance. The brain stem, the primitive portion of the brain, is responsible for reflexive or instinctive actions and connects the rest of the brain to the spinal cord.

Intelligent activity is generally the key to the mammals success and implies a certain degree of freedom of action. With their superior brains, mammals were able to compete successfully with animals that were much stronger, but less intelligent. When the dinosaurs were gone, the mammals competed with each other in a challenging environment, becoming even more brainy, adaptable, and varied.

The brain size of fossil animals can be determined directly by measuring the volume of casts, called *endocasts*, which are made from the animal's skull and are replicas of the cranial cavity. The study of brain size relative to body size indicates that living reptiles and other lower vertebrates have not departed significantly from their ancestors and have adapted without major advances in relative brain size. Neither were dinosaurs all that small-brained and were normal with respect to the relative brain size of reptiles. The brains of birds and mammals are more highly *encephalized*, meaning there is an additional amount of brain tissue in excess of what is required for simple transmission of impulses back and forth between the brain and the body. Therefore, they are higher vertebrates because they are higher in a scale of biological intelligence than reptiles and lower vertebrates.

There was a fourfold increase in relative brain size from the reptiles to the archaic mammals and then no substantial increase for at least 100 million years, indicating an adaptation to a lengthy, stable ecological niche during the Mesozoic. It was not until about 50 million years ago that there was another fourfold increase in relative brain size in response to the adaptive radiation of mammals into new environments since then, there has been steady advancement into the present.

When a mammalian species entered a new environment, its relative brain size increased rapidly to cope with the new conditions, and then its size was maintained for as long as the environment remained stable. For example, dolphins of 20 million years ago had obtained the level of intelligence comparable with living species such as the harbor porpoise, probably because of the stability of their ocean environment.

The ancient mammals, having descended from the mammallike reptiles that were driven into extinction by the dinosaurs some 150 million years ago, were forced into a nocturnal life-style that required the evolution of highly acute senses along with an enlarged brain to process the information taken in by those senses. This theory would account for the sudden jump in relative brain size and then stasis for the next 100 million years. The early mammals became more aware of their environment, sensing stimuli from a variety of sources and developing a nervous system for handling enormous amounts of incoming information. With the extinction of the dinosaurs and other ruling reptiles, the mammals became the recipients of daytime niches, along with a preponderance of new sensory signals for the brain to sort out, causing another rapid increase in relative brain size. The mammals also became more selective and flexible in their responses to outside stimuli; in other words, they were capable of making decisions.

It is generally thought that learning is the hallmark of intelligence and that animals are not truly intelligent, a characteristic reserved for humans alone. Yet, animals are capable of choosing among alternatives and seem to know in advance the outcome of those decisions. Learning is often thought to be an alternative to *instinct*, which is information passed on genetically from one generation to the next. Birds, for instance, seem to have fixed-action patterns like singing, derived from information implanted in their brains by their parent's genes. However, birds also can perform certain complicated learned tasks.

It appears that the process of learning in higher animals is guided by information inherent in their genetic makeup. In other words, learning is often controlled by instinct, and animals are preprogrammed to learn certain things in a certain manner. Therefore, animal behavior is mostly instinctive behavior tempered with learning. Certain behaviors are more easily learned than others because they are more natural, and although rats can learn to press a lever to obtain food, they cannot do the same trick in order to avoid an electric shock. Therefore, in certain behavioral situations, animals are innately predisposed to learn some things more readily than others.

Animals must learn a great deal about their environment in order to survive. Some they learn by trial and error, which is dependent on innate guidance, and some they learn by mimicking their parents. In many ways animals are smart, especially in those ways that natural selection has favored, and in some ways humans can be embarrassingly stupid.

THE WATER MAMMALS

Just as the lobe-finned fish that gave rise to the amphibians left the ocean to seek opportunities on land some 350 million years ago, so did the mammals seek similar opportunities, only this time in the opposite direction. Mesozoic relatives of the crocodiles, the extinct mesosuchians, practiced a sort of reverse evolution and became totally adapted to a predator's life in the sea.

Actually, there is no such thing as reverse evolution. Evolution can go in only one direction: from simple to complex (FIG. 6-14), and no organism has ever been shown to degenerate to a lower form. No one knows why the mammals entered the ocean to compete with the fish, but generally strong selective pressures, such as abundance of food and

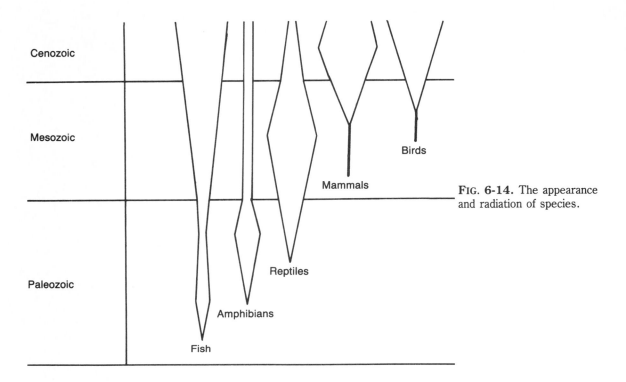

Cenozoic

Mesozoic

Paleozoic

Birds

Mammals

Reptiles

Amphibians

Fish

FIG. 6-14. The appearance and radiation of species.

habitats or intense competition, will drive a species into a new environment. Before this can happen, however, a species must be equipped to survive in its new environment or at least find it easier to live in. The lobe-finned fish developed the proper appendages, which enabled it to ''swim'' across the land. The archaeopteryx, although it could not fly, was at least on the right track toward modern birds.

Having been driven into the air by the dinosaurs and kept there by hunting mammals, birds found life a lot easier on the ground once the threat was eliminated because they must expend huge amounts of energy to stay aloft. Birds also have successfully adapted to a life in the water. Certain diving ducks are specially equipped for ''flying'' under water to catch fish. Penguins are large flightless birds that have taken to life in the water and are also well adapted to survive on the Antarctic ice cap in winter. Sea otters, manatees, seals, and walruses are not fully adapted for a continuous life at sea and have retained many of their terrestrial characteristics (FIG. 6-15).

There are some 70 species of entirely marine mammals that play just as important a role in oceanic ecosystems as mammals do on land. These mammals belong to the order *Cetacea*, which includes whales, dolphins, and porpoises. They are plentiful in all the oceans from the tropics to both polar ice caps, and many species migrate seasonally between cold and warm seas. The giant blue whale (FIG. 6-16) is the largest living animal on Earth and even dwarfs the largest dinosaur that ever lived. Because

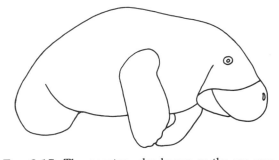

FIG. 6-15. The manatee, also known as the sea cow.

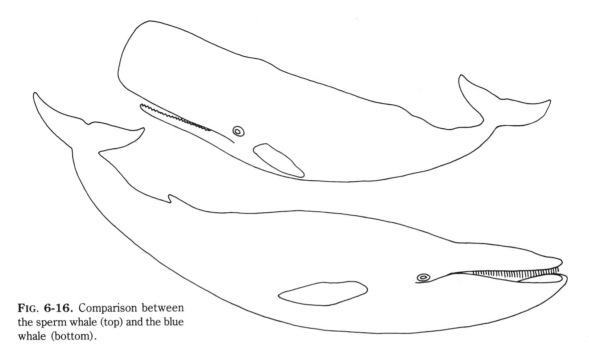

FIG. 6-16. Comparison between the sperm whale (top) and the blue whale (bottom).

cetaceans are warm-blooded and water removes body heat faster than air, they needed to take special measures to cope with the cold, which cold-blooded fishes found unnecessary. The cetaceans accomplished this with a thick insulating blanket of blubber, which nineteenth century whalers exploited almost to the extinction of certain species of whales.

The smallest cetaceans, because of their larger surface-area-to-volume ratio, are faced with the greatest problem of maintaining their body temperature in cold water. With a normal mammalian metabolic heat supply, a small porpoise could not possibly keep warm in northern waters. Regardless of their size, however, cetaceans maintain a normal internal temperature of about 98.6 degrees Fahrenheit, the same as for humans. They are able to maintain this temperature by having a basal metabolism rate that is about three times higher than terrestrial mammals of the same weight. As a result, they need to consume three times more food and three times more oxygen than do their terrestrial counterparts.

Whales, on the other hand, because of their large volume-to-surface-area ratio and their thick coat of blubber, have the opposite problem: that of overheating even in the coldest water. They rid themselves of excess heat by increasing the blood flow to the surface, thus, providing an effective heat exchanger. The lower basal metabolic rates of large whales allows them to feed off the store of fat in their blubber for long stretches while migrating through thousands of miles of barren waters. As a result, large whales are relatively independent of local food supplies, and they can satisfy their energy needs for up to six months by living off their blubber alone. This gives them the freedom to feed in the rich polar seas and travel thousands of miles to give birth to their young in warm tropical waters.

Whales also have the astounding ability of holding their breaths for exceptionally long periods, and sperm-whale dives can last for as long as an hour. They accomplish this feat by having a powerful heart and a huge lung capacity—up to three times more blood per unit of body weight than terrestrial mammals—and the ability to store large quantities of oxygen in their muscles. On the surface, the animal is able to rapidly exchange the air in its lungs, recharge its blood and muscles with new oxygen, and be ready to dive again. Whales can also dive deep in excess of 1000 feet and appear to be un-effected by the *bends*, a buildup of nitrogen in the blood that can be fatal to human divers.

In early October 1984, nearly 100 pilot whales were found lying dead or dying on the shores of Cape Cod, Massachusetts. Every year, for unknown reasons, whales beach themselves on the shores from Cape Canaveral, Florida, to Cape Cod. There is strong evidence suggesting that whales and dolphins navigate by using some sophisticated magnetic detector that senses changes in the Earth's magnetic field. Other animals known to possess a magnetic sense include certain bacteria, bees, fish, and migratory birds. Researchers have found that some ceta-ceans have little pockets of ferrous oxide in their heads. Like iron filings, which align themselves to a toy magnet, these deposits respond much the same way to the Earth's magnetism.

There was a surprisingly high coincidence of whale-stranding sites with *magnetic minima*, or areas where the Earth's magnetic field is at a local minimum. Apparently, these marine mammals use magnetic minima to navigate long distances at sea. The minima often form long "stripes" in a north-south direction on the ocean floor, and cetaceans following them might become disoriented when entering a magnetically anomalous area. If the whales are not returned to the water soon, they suffocate be-

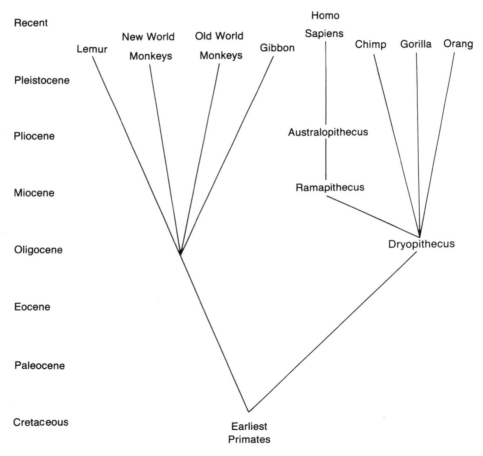

FIG. 6-17. The primate family tree.

cause of the weight of their three-ton bodies or die from shock caused by a breakdown of their thermoregulatory system.

THE PRIME OF LIFE

Primates evolved from a small squirrellike mammal some 50 million years ago. Thereafter, the primate family tree split into two branches: the monkeys, and the apes, which include the hominoid line (FIG. 6-17).

Apparently, relatively little evolutionary change has taken place in the monkeys since the Middle Miocene, about 20 million years ago. The common ancestor of the New World monkeys probably migrated from Africa to South America prior to 30 million years ago when the two continents were still close enough together. From then until about 25 million years ago, they evolved rapidly apart from the Old World monkeys.

Apes appear to have evolved along more complicated family trees with a greater diversity along hominoid lines. Approximately 30 million years ago, the precursors of modern humans and apes lived most of their lives in the trees of the dense tropical rain forests of Egypt, which is now mostly desert. These apelike ancestors spread from Africa into Europe and Asia during the Miocene epoch from about 25 to 10 million years ago.

A link between African and European apes is believed to have lived on a group of islands near northern Italy. However, it is still a mystery how an ape with long arms and short legs could swim such a great distance.

The use of tools was once thought to be an entirely human attribute. Yet many animals use tools, including apes, otters, birds, and even wasps. Chimpanzees have been known to peel the leaves off a twig, stick it into the opening of a termite's nest, and withdraw a tasty morsel. The most adaptive and inventive simian tool user in captivity is the orangutan, which has been observed using a hammerstone to make a sharp stone flake in order to cut the string around a box of food.

Not even chimps have been observed to use a tool to make a tool, but neither have orangutans been observed to use tools in the wild uninfluenced by humans. Gorillas, which are closely related to humans, and chimps have only infrequently been seen to use tools in captivity and never in the wild. Even gibbons have been known to hurl branches at humans standing below their trees. Apparently all great apes are smart enough to use tools, but they only do so in useful circumstances. Nevertheless, the extent and nature of tool use among apes remains a subject of considerable importance in relation to the development of tool making among the earliest members of the human family.

7

The Big Brain

GENERALLY, it is thought that what separates humans from the rest of the animal world is our big brains. More correctly, it was the adaption toward permanent bipedal walking and running, which set the hands free for tool-making and other useful purposes, that led to human evolution.

The conversion from a four-legged to a two-legged mode of travel came at a price, however. The *foramen magnum*, the hole in the skull through which the spinal cord passes, had to face directly downward, not come out of the back of the skull as in four-footed animals. The skull had to be balanced on top of the spinal column, which needed to support the body weight in a vertical, rather than a horizontal, posture, generally accompanied by back problems. The legs had to be very long in relation to the trunk and strong enough to bear the full body weight. The foot had to have longitudinal and lateral arches for softening the impact on the ground. The big toe had to point forward parallel to the other toes in order to take up much of the stress during walking, as well as balancing control.

In 1976, well-preserved footprints of two of our ancestors were found embedded in a volcanic ash bed at Laetoli, Tanzania. The footprints appeared remarkably modern, with rounded heels and arches, pronounced balls, and forward-pointing toes—all the features necessary for walking erect. The ash bed was 3.75 million years old, making the footprints the earliest evidence that humans walked upright during that time.

There must have been strong selection pressure for humans to take to walking on two legs. Bipedalism could have developed for long-distance migration in response to migrating herds. Early hominids were probably highly mobile scavengers, rather than hunters. The hands would then be free, not for carrying weapons, but for carrying children. Therefore, bipedal walking became a more efficient way of carrying offspring than the method chimpanzees or baboons must use to carry their young as they move on four feet through relatively open country. A migratory scavenger's mostly meat diet would work toward natural selection of a powerful striding gait because hunting would be a much more demanding way of life than foraging, bringing about evolutionary changes in the hip and foot. Direct competition with and danger from better equipped four-legged scavengers such as hyenas would have provided selection pressure for the skill to fashion stone tools with which to break through the tough hide the teeth were poorly designed for. Walking upright with the head held high also let early humans see over the tall savanna grass on the lookout for predators.

HUMAN HERITAGE

Human beings belong to the order Primates, which also includes apes, monkeys, and prosimians such as lemurs. Within this order is the superfamily Hominoidea, which includes only humans and apes. Within that superfamily is the family Hominidae, which includes only humans and their extinct human-like ancestors. African apes, including chimpanzees and gorillas, are more closely related to humans today than the Asian apes like the orangutans and gibbons.

The radiation of all the great apes from an ancestral stock took place in the Middle Miocene epoch around 15 million years ago, with the hominids splitting from an African ape lineage by the Late Miocene. The human line diverged from the hominoids 7 to 5 million years ago, much later than it was previously thought. Nor did the hominids and great apes share a last common ancestor after the lineage leading to humans diverged; in other words, no apes evolved from the human line. Therefore, features common to all apes and hominids like the absence of a tail were either retained from the last ancestor of all large hominoids or evolved independently in different species.

The earlier large hominoids were much more diverse, whereas today, only five genera have survived. The best known of these Miocene apes was called Proconsul which was a small animal about the size of a baboon and once thought to be the ancestor to all hominoids. However, some of its body features were unlike those of any living higher primates, and many of its other features were also unique.

Therefore, it was probably just a side branch, leading nowhere on the primate family tree.

The split between African and Asian apes most likely coincided with the linkup of Africa-Arabia with Eurasia in the Middle Miocene some 17 million years ago, as the result of continental drift. This linkup allowed the migration of African hominoids along with other mammals into the rest of the Old World (FIG. 7-1).

The continental collision raised mountain chains, caused the Tethys Sea to disappear, and produced a shift in oceanic and atmospheric circulation patterns, which in turn brought about changes in climate and habitats. From the Middle to Late Miocene, the Eurasian climate was seasonal, with winters much milder than they are today. What is now grassland and desert was then woodland, and the forests were much more widespread.

The early Asian hominoids were represented by *Ramapithecus*, which differs from other Miocene hominoids in certain characteristics that resemble those of later hominoids. These features led to previous speculation that *Ramapithecus* was an early hominid, and that the hominids diverged from the African hominoids at least 15 million years ago. It now appears that *Ramapithecus* was more closely related to the sole surviving Asian great ape, the orangutan. From 14 to 4 million years ago, there lies a large gap in the fossil record, which jumps from the hominidlike but mainly ape form of *Ramapithecus* to the true hominids.

The early hominids called *Australopithecus* were markedly different from any living today. They first appeared perhaps 4 million years ago in Tanzania and Ethiopia and as much as 5 million years ago in northern Kenya (FIG. 7-2). One remarkable discovery of an almost half complete skeleton of a female *Australopithecus* found at Hadar, Ethiopia, was named ''Lucy'' and dated between 3 and 4 million years old.

These hominids were primitive in most of their features. They were generally from 3 to 4 feet tall and weighed upwards of 100 pounds, with males up to twice as large as females. They had faces resembling chimpanzees, teeth with broad molars and small

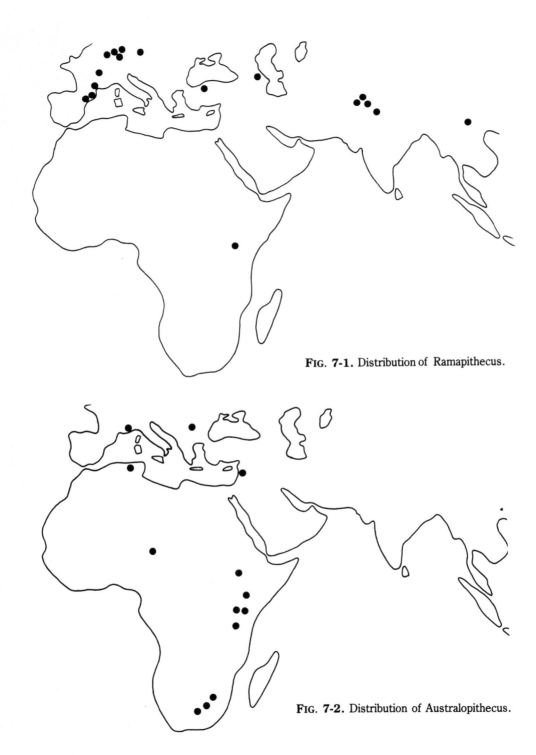

FIG. 7-1. Distribution of Ramapithecus.

FIG. 7-2. Distribution of Australopithecus.

canines, skulls like gorillas (FIG. 7-3), and brains the size of living African great apes—around 400 cubic centimeters. Features of the hip, knee, ankle joints, and foot indicated that these early hominids were clearly bipedal, but the arms were relatively long and the legs were relatively short. Their hands were capable of powerful grasping and probably more manipulative than those of living chimpanzees. Lucy and her relatives lived in an area of woodland and savanna away from the great Pliocene forests. They probably had not acquired the skill of tool-making, for no stones altered into tools have been found associated with the hominids bones. Instead, they might have used similar kinds of temporary tools that living chimpanzees use, such as wood and stones, adapted for food gathering and processing.

Around 2.5 million years ago, the climate became cooler, causing a shift in Africa toward more open, savannalike habitats. This shift resulted in the appearance of many new animals and spurred the evolution of the hominids. At this time, there were probably three or more species of African hominids. Robust (heavy-boned), small-brained hominids lived in the same area as gracile (light-boned), large-brained hominids, and it is possible that one form preyed upon the other. The largest, called *Gigantropithecus*, was larger than a gorilla, estimated at over 6 feet tall, weighed nearly 400 pounds, and was an evolutionary dead end, becoming extinct about 0.5 million years ago.

Another large species of robust *Australopithecus* from southern Africa was small-brained, and the male was markedly larger than the female, a condition called *sexual dimorphism*. The exaggerated size of males compared to females probably resulted from sharp competition among males for females. With more cooperation among males, sexual dimorphism diminished.

These hominids appeared more than 2 million years ago and survived practically unchanged for 1 million years. They might have dug with sticks and hammered with stones, but there is no evidence that they used tools to any large extent. After 1 million years of apparent stability, for unknown reasons (possibly a change in climate or habitat), this animal, too, became extinct. The gracile Australopithecines

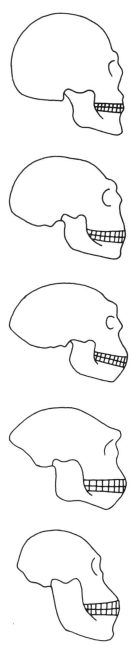

FIG. 7-3. Comparison of human skulls. From bottom to top: Ramapithecus, Australopithecus, Homo habilis, Homo erectus, Homo sapiens.

represented by Lucy and a well-preserved 2 million-year-old skull of a 6-year-old infant named the "Taung child" gave rise to the *Homo* line of hominids.

THE HANDY MAN

From the Australopithecine line a little more than 2 million years ago arose a larger brained hominid called *Homo habilis*, named by its discoverers, Louis Leakey and his coworkers, in Kenya in 1964. It was like *Australopithecus* in many ways, such as its face and teeth. However, it had a significantly larger brain, which averaged about 700 cubic centimeters—about half the size of modern human brains. The limb bones were unlike those of *Australopithecus* and resembled later species of the genus *Homo*. The resemblance probably reflected changes in the mode of locomotion and in the dimensions of the female pelvis, demanded by the size of the newborn infant, which required a long gestation period.

Along with the appearance of *Homo habilis* came the first archaeological sites (FIG. 7-4), consisting of concentrations of used or altered stone, often brought from some distance away, along with animal bones. The sites show a shift in diet to include more animal food, which might represent a link between dietary change and increased brain size. It is probable that *Homo habilis* was a hunter-gatherer, transporting food to a home base to be shared with others, and that he practiced a division of labor, leaving the hunting to the males and the food-gathering and child-rearing to the females.

Because *Homo habilis* was considered an evolutionary intermediate between the relatively small *Australopithecus* and the relatively large *Homo erectus*, it was assumed that it would also be of intermediate size. A partial skeleton of an adult female *Homo habilis* found at Olduvai Gorge, Tanzania, in 1986 that was 1.8 million years old was not much taller than Lucy who, although over 1 million years older, stood a little over 3 feet tall. Apparently, human evolution did not take place through gradualistic changes, but was abrupt, with a large modification in body form between *Homo habilis* and *Homo erectus*. This meant the primitive body form that

characterized the earliest hominids continued much later in human history.

The extreme sexual dimorphism in which the males were twice as large as the females also continued unabated through *Homo habilis* and diminished only with the advent of *Homo erectus*. Different degrees of sexual dimorphism are usually associated with different social systems; therefore, the first member of the *Homo* genus seemed less human and more apelike. It was more like a mosaic of primitive and derived features, and although it walked bipedally, it seemed to retain the hominoid capacity to climb trees (although it did not necessarily do so and the resemblance could just be a genetic carryover). Like Lucy, *Homo habilis* had arms nearly the same length as its legs, but the foot was more humanlike.

Homo habilis was probably a part-time scavenger with a diet of meat supported with fruits and vegetables. Also, the pattern of wearing on the teeth seems to indicate a diet that was mostly fruit. The animal lacked the speed to be a successful full-time hunter or the body size to compete successfully on a full-time basis with fierce scavengers. *Homo habilis* lived in open country or wooded environments, requiring it to cover wide areas in search for food. Bipedalism with a strolling pace sustainable for long periods would be suitable for a scavenging lifestyle that also required tool use. Compared with other meat-eaters, *Homo habilis* was poorly equipped for tearing flesh from a carcass without the aid of a sharp implement of some kind. Crude flakes made of flint for cutting and cobbles used as hammerstones might have made up the toolkit of *Homo habilis*.

The transition between *Homo habilis* and *Homo erectus* and from scavenging to a hunting way of life appears to have taken place between 1.8 and 1.6 million years ago. Brain size increased further, and the anatomical adaptation to upright walking became increasingly advanced. What ever became of *homo habilis* is unknown, for it disappeared from the African plain 1.75 million years ago. It survived for just a few hundred thousand years, only to be replaced by a much more durable and larger brained species: *Homo erectus*.

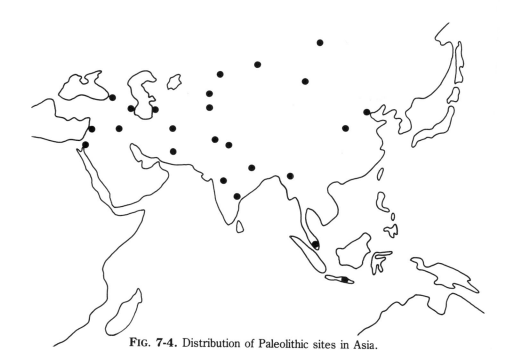

FIG. 7-4. Distribution of Paleolithic sites in Asia.

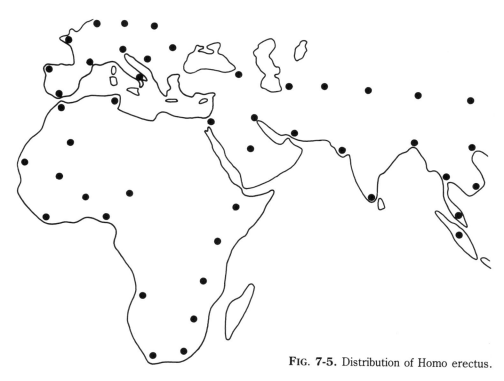

FIG. 7-5. Distribution of Homo erectus.

THE ERECT MAN

It is generally believed that *Homo erectus* was a direct descendent of *Homo habilis*. It was the first widely distributed hominid (FIG. 7-5), appearing in Africa 1.6 million years ago. By 1 million years ago, the species was present in southern and eastern Asia where it lived until a little over 200,000 years ago. It has even been suggested that *Homo erectus* originated in Asia and migrated to Africa, implying that it evolved independently of *Homo habilis*.

During a time span of well over a million years, the body form as well as behavior of *Homo erectus* showed a prolonged stability. *Homo erectus* more closely resembled later species than it did earlier ones, with a larger, more robust body and a larger brain—800 to 1100 cubic centimeters.

Note that brain size is not an all-conclusive indicator of human intelligence, and that a hominid with a larger brain case was not necessarily more intelligent. Nevertheless, the gradual increase in cranial capacity generally aids in placing species in their proper sequence in the *Homo* evolutionary line.

The front teeth of *Homo erectus* resembled those of earlier hominids, but the cheek teeth and face were smaller, and the size of the brain case seemed to overwhelm the mouth, as it does in living humans. The skull differed markedly from the modern form; it was much thicker and flatter, and had protruding brow bones. The species was better adapted for bipedal locomotion (hence its name), with a skeleton much like that of modern humans, except the bones had thicker walls and narrower marrow cavities.

Homo erectus was close to modern humans in average height—between 5 and 5.5 feet tall—and probably weighed a little over 100 pounds. The men might have lived to be about 30 years old, and the women probably outlived the men because they were less exposed to danger and mortal injury from hunting big game, although there might also have been a high percentage of female mortality from childbirth. There was probably still considerable sexual dimorphism among earlier species, but this diminished as males began to cooperate more in hunting.

Homo erectus lived contemporaneously with other lower forms of hominids and might have even hunted them, possibly to their extinction. Some members of *Homo erectus* made larger, more competent flaked stone tools or hand axes. Some later populations might have used more sophisticated methods of producing and modifying stone tools. They developed quite an elaborate culture, characterized by inhabiting caves and hunting game. Communication was probably not much more than grunts, gestures, and facial expressions similar to those used by living apes. It became necessary to communicate in order to organize the hunt and to pass down tool-making skills from one generation to the next. Some populations of *Homo erectus* might have been the first to use fire, possibly for hunting, cooking, and keeping warm.

A variety of *Homo erectus* called "Peking man" lived in a cave at Zhoukoudian about 30 miles southwest of Beijing (Peking), China continuously for nearly a quarter million years, starting about half a million years ago. Fossilized animal bones indicated that Peking man was an effective hunter of both large and small game, and by this time, he was able to compete successfully with large carnivores.

With his use of weapons, Peking man was able to go beyond the limits of his size and successfully hunted big game, such as deer, that were both larger and faster than he was. Fruits and seeds were also a large part of his diet, as indicated by fossilized seeds in the cave. An abundance of flaked-stone implements indicated a high degree of tool-making skill. Quartz and flint were carried from some distance away and shaped by using another stone into various implements, including scrapers, choppers, awls, and points.

Peking man had the ability to control fire and keep it burning for some time, although it is not certain whether he could ignite a fire and probably had to rely on natural fires set by lightning. Since Peking man lived in a temperate climate, the use of fire became important for survival in the cold winters. There is also every indication that he used fire to cook his food, as evidenced by quantities of charred seeds found in the cave.

The rigors of hunting were probably too difficult for women, who would otherwise be preoccupied with child bearing and rearing and gathering fruits and seeds. This established a division of labor that is still common among hunter-gatherer societies today.

THE CAVE MAN

The first *Homo sapiens* included the Neanderthals, named for the Neander Valley in Germany where the first fossils were recognized in 1856. Although further up on the evolutionary ladder toward modern humans, the Neanderthals were probably as unlike us behaviorally as they were physically. Their skeletons were much more heavily built, and muscle attachments on the bones indicated that they were much stronger than we are. The skulls had large, bony brow ridges, a large face, and a long, low brain case, probably derived from *Homo erectus*. However, the cranium held a brain that was a little larger than that of modern humans. A larger brain only reflects the fact that Neanderthals were bulkier than us and does not necessarily mean they were more intelligent.

Earlier bone reconstructions showed the animal had a stooped, apelike appearance, which led to the erroneous assumption that the Neanderthal was greatly inferior to modern humans. Their teeth were larger and heavily worn, an indication that they were used for a variety of nonfeeding tasks such as chewing animal hides to make them softer for use in clothing and shelter (in the form of skin tents). The Neanderthals were generally thought to shelter in caves because that is where most of their bones were found, although this is because caves preserved bones better than open sites.

The Neanderthals ranged from western Europe to central Asia (FIG. 7-6), beginning about 130,000 years ago. They lasted until about 35,000 years ago, and they were able to endure the rigors of the cold climate during the last ice age. The activity of the Neanderthal's massive muscles could have supplied their bulky bodies with more heat. Other physical attributes might also have been adaptations to the cold, subarctic conditions such as those of modern Eskimos, but Neanderthals in more temperate regions were nearly equally equipped.

Although some Neanderthals occupied caves, others lived in the open, as indicated by open-air sites with hearths, rings of mammoth bones, and masses of stone tools normally associated with Neanderthals. Most of these tools were flakes of flint that were struck from a core stone and trimmed into projectile points, knives, and scrapers. This assortment of tools was much the same everywhere and was called the *Mousterian* or middle Paleolithic, *industry*, named after a site in Le Moustier, France, where such tools were first discovered.

Although the Neanderthals made Mousterian tools, not all Mousterian tool-makers were Neanderthals. It is a general level of tool manufacturing, and not a distinct expression of Neanderthal intellect and skill.

Their way of life probably did not differ radically from recent hunting peoples. They formed into hunting bands that were probably linked loosely into tribal groupings or groups with a common language.

Despite their primitive physical appearances, there is every indication that the Neanderthals were becoming culturally refined peoples by about 40,000 years ago. Although Neanderthals' hands were larger and had a much stronger grip than those of modern humans, there was nothing gorillalike about them, and the control of movement was evidently as good as ours. Their tool-making skills were every bit as good as those of anatomically modern peoples living during the same time. The Neanderthals might have even made rock carvings and cave paintings. They certainly buried their dead and placed grave offerings such as ibex horns or flowers with them. They might have conducted these burial rituals with a belief in an afterlife, worshiped or revered animals such as the deer and the cave bear, and performed other ritualistic activities, possibly including cannibalism.

By about 35,000 years ago, the Neanderthals were abruptly replaced by more anatomically modern humans. Their sudden departure after some 100,000 years of prosperity might have been the result of some form of social degradation or tribal warfare that left a vacuum for more modern humans to

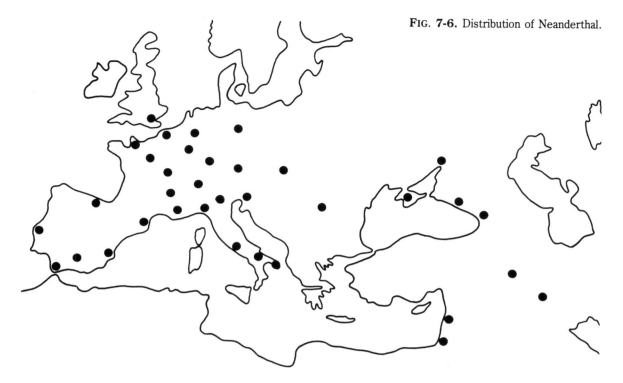

FIG. 7-6. Distribution of Neanderthal.

fill. It seems highly unlikely that the Neanderthals evolved quickly into modern human beings. Interbreeding with advanced peoples living at the same time, if it were possible, would have produced intermediate body forms, sort of like breeding a donkey with a horse, producing an infertile mule. Nevertheless, no transition between Neanderthals and modern humans has been convincingly demonstrated, nor does it seem likely that a race with superior hunting skills decimated them in a sort of Stone Age holocaust.

MODERN MAN

It is not certain where exactly modern humans sprang from. Perhaps it was as early as 60,000 years ago in southern Africa or the sub-Saharan where some of the oldest finds of modern man were discovered. They were called the Cro-Magnons, *Homo sapiens sapiens*, named after the Cro-Magnon cave in France where the first discoveries were made in 1868. They were unlike the Neanderthals and shared the great majority of physical attributes of humans today. Their cranial capacity was as large or larger than that of recent humans, and their brain-case proportions were modern, rather than Neanderthal. The skull was short, high, and rounded without the large brow ridges; the face was robust and flat like that of modern Europeans; and the lower jaw ended in a definite chin.

Cro-Magnon skeletons were slender and long-limbed, in contrast with the stocky Neanderthals. The main selective forces that favored the modern physique over the Neanderthal might have been climate change or cultural advance, whereby improved stoneworking techniques and associated behavioral changes might have given a significant adaptive advantage to the less heavily bodied people.

At some time during the last glacial period, between 45,000 and 35,000 years ago, Cro-Magnon made his advance into Europe and Asia, probably during an interlude when the ice age climate was not so severe (FIG. 7-7). The European geography at this time was largely composed of grasslands, rather

FIG. 7-7. Distribution of Cro-Magnons in northern Europe.

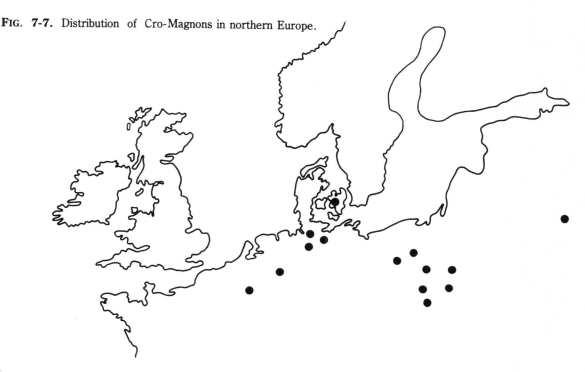

than tundra. On this range roamed reindeer, wooly mammoths, and other cold-adapted species, and Cro-Magnon hunted and carved up the carcasses with finely honed blades. Deadly spears, often with animals carved or engraved on them, became a most effective weapon for hunting big game.

Music assumed an important role in Cro-Magnon culture during the Upper Paleolithic from 35,000 to 10,000 years ago. The first known instrument was a bone flute found in France dating from about 30,000 years ago. By about 23,000 years ago, sewing needles made of bone appeared in southwestern France, and they provided a more sophisticated means of tailoring cold-weather clothes. One of the most important developments was the trading of goods across the countryside, especially body ornaments such as seashells used for making necklaces, which were found at sites up to 100 miles from the seacoast. Jewelry was also made from stone, ivory beads, and the teeth of dangerous animals such as bears and lions.

Cave paintings were another form of human expression, and one cave in the French Pyrenees has walls containing over 200 human handprints, mostly with missing fingers, dating about 26,000 years ago. The fingers might have been hacked off in some sort of ritual or were destroyed by disease and infection. The best known art objects were the so-called Venus figures, dating around 26,000 years ago. They were apparently small, elaborate sculptures of fertility symbols. The late ice-age people tended to make elaborate and beautiful drawings and carvings of animals, especially those they did not eat, such as horses, probably as some sort of ritual.

The Cro-Magnons became prolific and widespread during the ice age, ranging throughout Europe and Asia, despite the advancing ice sheets. What attracted these people into these desolate, frozen areas was a rich stock of large animals that were adapted to the cold. These ice-age peoples probably lived much like present-day Eskimos and Lapps, fishing the rivers, possibly using small boats, and hunting reindeer and other animals both large and small, including birds. Lacking wood in the cold tundra to build their homes, ice-age hunters of the central Russian plain used mammoth bones and tusks covered with animal hides. They burned animal fat to keep warm. It is possible that ice-age man over-hunted the mammoth and some other large animals, causing them to become extinct. A changing climate to our present interglacial also could have caused the extinctions.

By crossing a land bridge between Siberia and Alaska across the Bering Strait, ice-age peoples were able to populate the Americas (FIG. 7-8) as early as 32,000 years ago. The massive continental glaciers locked up great quantities of water, which substantially lowered the sea level and exposed parts of the shallow seafloor to the surface. From North America, these early Americans crossed over the Panama isthmus into South America and roamed as far south as southern Chile. In addition, cave paintings in Brazil suggest that cave art began in the Americas about the same time it appeared in Europe and Africa.

The peopling of the Pacific around 40,000 years ago was the greatest feat of maritime colonization in human history. Human populations of modern anatomical form occupied the continent of Australia 32,000 years ago and probably 8,000 years earlier than that. By island-hopping from southeast Asia, people managed to settle the great ocean reaches, and it is even contended that they went as far east as South America, although a northern route through North America seems more plausible.

The long climb up the human evolutionary ladder was not without its pitfalls, dead ends, and setbacks. However, the Upper Paleolithic saw more advancements in the human condition in 25,000 years than in the entire 2 million years since stone

FIG. 7-8. Ruins of Anasazi Indians, Mesa Verde National Park, Colorado.

(Courtesy of National Park Service)

tools were first employed. The Neolithic period, which began 10,000 years ago after the ice sheets retreated near their present positions, was a time of unprecedented human achievement.

The dawn of agriculture at the beginning of the Neolithic has been said to be the greatest achievement and the worst mistake in the history of the human race. Although it freed modern societies from the constant search for food, giving people time to contemplate other aspects of life, it also made people territorial, by allowing them to settle permanently in one place. Great armies were raised to capture other people's territories, and the blood bath continues into the present. With agriculture came gross social inequalities, corruption, disease, and despotism. The freedoms of the simple hunting and gathering way of life gave way to a complex culture and slavery, with the less fortunate toiling long, hard hours in the fields to feed the nonproductive elite. Population growth, which was previously held in check by the harsh demands of nature, soared when nature came under the plow. In just 10,000 years, the number of humans increased geometrically from about 5 million to 5 billion today. Overpopulation is causing greater hardships than all the trials and tribulations man has faced in his entire 5 million-year existence on Earth.

8

The Life Cycles

JUST about every aspect of life on Earth is governed by cycles of some sort, and it was probably the periodicity of the Sun-Earth-Moon system (FIG. 8-1) among other things that gave Earth life in the first place. The Moon raises tides in the ocean, in the atmosphere, and even in the crust. The constant waning and waxing of the tides are responsible for the prodigious growth in the intertidal zones. The pull of the Moon has been blamed for everything from drought to lunacy.

The solar cycle appears to have an effect on the climate, which in turn affects life on Earth and which can be detected in tree rings. This cycle has existed since Precambrian time and is thought to be responsible for cold periods and drought.

The Earth's orbital cycles have a large effect on the climate. As the Earth's orbit takes it farther from the Sun during one season and closer the next, the amount of sunlight the Earth receives changes. This process might be responsible for the ice ages. The cyclic changes in the tilt of the Earth's axis determine the amount of solar input by latitude. The precession of the equinoxes reverses the seasons

on a cyclical basis, making it winter when before it was summer. The Earth has its own inner cycles, which affect geological and biological processes and which might be influenced by outside forces such as the gravitational pull of the Sun, Moon, and other planets.

THE SOLAR CYCLE

The study of the Sun and its influence on the Earth is one of the greatest challenges to astrophysicists. A systematic observation of sunspots has been going on since Galileo first spotted them through his telescope in the early seventeenth century. During the period from 1645 to 1715, there was apparently a minimum of sunspot activity (FIG. 8-2), known as the *Maunder Minimum* for the English astronomer Walter Maunder, who discovered it in 1894. It was blamed for a span of unusually cold weather in Europe and North America, during what was called the ''Little Ice Age'' from the sixteenth to the early nineteenth century.

In 1843, the German astronomer Heinrich Schwabe pointed out the existence of a pronounced

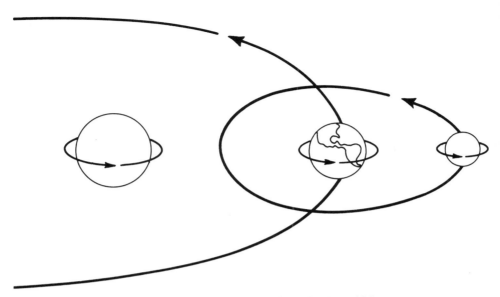

FIG. 8-1. Orbital motions of the Sun, Earth, and Moon.

| 1700 | 1800 | 1900 |

FIG. 8-2. Sunspot peaks through time.

cycle of sunspot activity approximately every 11 years. In the course of each cycle, sunspots, which individually last for only a short while, tend to initially appear at high solar latitudes then migrate progressively nearer the solar equator, showing that the equator was rotating faster than the poles. Other solar activity—including solar flares (FIG. 8-3), solar cosmic rays, ultraviolet rays, and X-rays—varies directly with the solar cycle. Moreover, the polarity of the Sun's magnetic field reverses every two sunspot cycles, or 22 years. The sunspots are associated with these magnetic fields, which are several thousand times stronger than the magnetic field on the Earth's surface.

The solar cycle might be the result of a global oscillation of the Sun, which rhythmically changes its diameter and results in variations in solar luminosity. The oscillation could be produced by a permanent internal magnetic field in the radiative interior that is coupled to an oscillating field in the turbulent convection zone having an intensity comparable to that of sunspots. The role of the turbulence in the convection zone is to concentrate the field at the base until it becomes unstable, which then signals the beginning of a new sunspot cycle. The zones of variation of the Sun's rotation rate might drive convective rolls in the convection zone. Gases carrying heat from the interior to the surface cool and then roll over as they descend to be reheated again, forming rolls similar to hair rolled on curlers. The Sun's core, which rotates faster than the rest of the Sun, also might partake in the cycle, with variations of angular velocity that oscillate with the solar cycle.

The oscillations seem to affect the climate on Earth. Studies of drought in North America have shown a drought period of 22 years, the same as the solar cycle. Apparently, the climatic variation results from the modulation of the Earth's ionosphere by charged particles from the Sun.

For centuries, scientists have spoken of a *solar constant*, and that the Sun has been shining steadily for eons. However, satellites monitoring the Sun have indicated that it has been slowly fading temporarily since the beginning of the 1980s. Satellite radiometers have measured a decrease of solar *irradiance*, or brightness, of nearly 0.1 percent from 1981 to 1984. A decrease in luminosity as small as this lasting for a decade could have an effect on global climate. During the Little Ice Age, the irradiance might have been reduced by as much as 1 percent, causing a drop of an estimated 2 degrees Fahrenheit in the average global temperature.

The apparent decrease in solar output might be caused by the decrease in the number of sunspots since the beginning of the decade. The sunspots themselves do not block out the Sun, as was once thought, but instead are an indication of increased solar activity. Sunspots appear dark because they obstruct the convective flow of heat toward the sur-

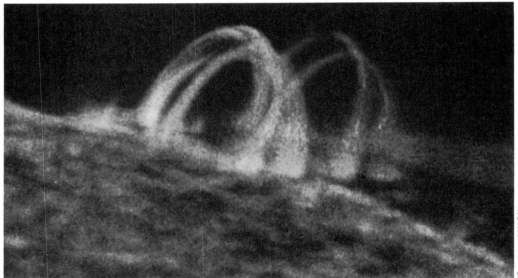

(Courtesy of NASA)

FIG. 8-3. Magnetic loops in solar active regions.

face; at the same time, they uncover the hotter depths of the Sun and thus increase their radiation into space.

The sunspot numbers peak about every 11 years and change their magnetic polarity about every 22 years. The 11-year cycle reached a minimum in 1975, a maximum in 1981, and a minimum again in the autumn of 1985. The Sun's irradiance should increase as each new cycle begins.

Claims that sunspot cycles have an influence on the weather are still surrounded by controversy because there is no clear evidence for such an influence. The best source for this information comes from a series of Precambrian glacial varves in Australian lake deposits, which date around 680 million years old. These varves are alternating layers of silt (FIG. 8-4) that were laid down annually in a lake below the outlet of a glacier during the late Precam-

brian ice age. Each summer when the glacial ice melted, meltwater that was turbid with sediments discharged into the lake, where the sediments settled out forming a banded deposit. During times of intense solar activity, the Earth's climatic temperature increased slightly, causing more glacial melting, a greater amount of discharge, and consequently, thicker varves.

By counting each silt layer, scientists have determined a continuous stratigraphic sequence spanning some 19,000 years. The sequence mimicked both the 11-year sunspot cycle and the 22-year solar cycle, and these similarities argue for a direct relationship between varve thickness and solar activity. A problem arises when trying to find similar periodicities in younger rocks, which are not as obvious as those in the Precambrian. The data does suggest, however, that solar activity has not

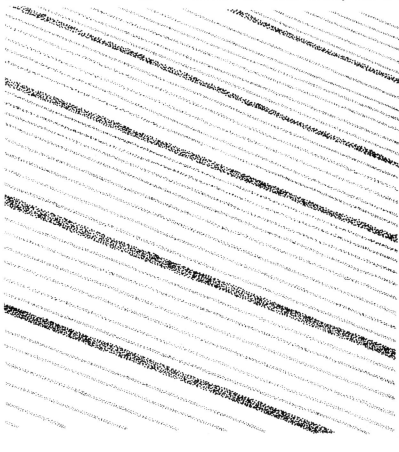

FIG. 8-4. Australian desert varves.

changed significantly over the past 680 million years, indicating an apparent stability of the Sun.

Tree rings (FIG. 8-5) also provide reliable evidence for a link between drought in the western United States and the 22-year solar cycle, the 18.6-year lunar cycle, or both. The tree ring data have been collected and used to reconstruct a regional series of annual precipitation for almost four centuries, starting from 1600. During a drought or an unusually cool season, a tree's growth can be stunted, which in turn results in a reduction of its ring width. Trees living between 1645 and 1715 gave an account of anomalous climatic activity that coincided exactly with the Maunder Minimum of sunspot activity during the Little Ice Age.

The tree rings are obtained by taking small core samples from living trees, such as the great sequoias of northern California (FIG. 8-6), which are among the world's oldest living plants. Also, by analyzing the radioactive carbon-14 content of rings from ancient, well-preserved trees, a history of the carbon-14 in the atmosphere has been reconstructed going back more than 7,000 years. The amount of carbon-14 generated in the atmosphere varies directly with the amount of solar cosmic rays, which are more abundant during times of intense solar activity. When they strike nitrogen atoms, the cosmic rays cause them to mutate into carbon-14 atoms. The carbon-14, along with ordinary carbon-12, is taken in by plants. When they die, the carbon-14 clock starts ticking away, providing a fairly accurate method of dating the wood. A continued study of tree rings could eventually provide important data for testing solar and lunar influence on the climate.

FIG. 8-5. Tree sample prepared for annual growth-ring studies.

(Photo by L.E. Jackson, Jr., courtesy of USGS)

FIG. 8-6. The General Sherman Tree, a giant sequoia that is one of the largest trees on Earth and estimated between 2,500 and 3,000 years old, Sequoia National Park, California.

(Courtesy of National Park Service)

THE LUNAR CYCLE

Evidence used previously to support a solar connection with drought has uncovered a lunar cycle whose effect on the climate, when combined with the solar cycle, might sometimes wreak such widespread climatic havoc as the great Dust Bowl of the 1930s. The effect of the Moon on the climate first was seen in tree rings, which were originally analyzed for the Sun's effect on drought in the Great Plains. The narrowing of tree rings during drought conditions revealed a tendency toward an expansion of the area of drought approximately every 22 years over a 360-year period.

The apparent effect of the 22-year solar cycle on the weather, however, was not as large as the 11-year cycle and accounted for only about 10 percent of the variance in the drought record. Therefore, the sunspots could not control drought, but only increase or decrease the tendency toward it. In addition, the effect was unpredictable, ranging anywhere from barely detectable to catastrophic. The larger effects on the climate might be a result more of the Moon's 18.6-year cycle of changing declination, whereby its maximum distance from the Earth varies by twenty times its diameter, or about 40,000 miles.

The effect of the Moon on the climate has been further confirmed in a study of the Indian monsoons, in which the Moon can accelerate or retard the seasonal march of the monsoon rains into India. The lunar cycle has also been detected in the record of floods and droughts in northern China going back 500 years.

The Moon's effect on precipitation in the Great Plains and elsewhere might result from gravity-induced tides in the atmosphere. The lunar cycle causes varying tides in the atmosphere, just as the daily motions of the Moon raises tides in the ocean. The atmospheric tides might produce a tide-induced wave in the atmosphere. Mountain ranges like the Rockies, which could have an influence on weather conditions in the Plains states, also might affect such a wave.

A similar atmospheric oscillation produces a wave of cloudiness high over the Indian Ocean every 40 to 50 days. As the wave sweeps eastward into the Pacific Ocean, it intensifies with speeds up to 20 miles per hour, and dies out when it reaches the eastern Pacific. As it circles the tropics, the oscillation can set other parts of the atmosphere—as far away as the poles—pulsating at nearly the same frequency. It is thought that this phenomenon might play a role in triggering the onset and withdrawal of the monsoons in India and cause the rains to pause in midseason. The effect also might trigger El Niños, which produce warm currents in the south Pacific that are detrimental to fish populations off the coast of South America. The atmospheric oscillation might even affect the jet stream, which plays a pivotal role in shaping the weather in North America.

Tides in the ocean (FIG. 8-7) are caused by the gravitational pull of the Moon and Sun and have a large influence on many aspects of life along the coast. The Moon revolves around the Earth in an elliptical orbit and raises the highest tides when it is closest to the Earth. The oceans flow into two tidal bulges, one facing toward the Moon and the other facing away from it. The tidal bulges follow the Moon around the Earth, taking about 27.5 days to make a complete circuit. Because of the rotation of the Earth, each point on the surface goes into and out of both tidal bulges once a day, causing most tides to rise and fall twice daily. Because the Moon orbits the Earth in the same direction as the Earth revolves, the tidal bulges move ahead a little each day so that the period between high tides is about 12.5 hours. Some areas might have only one tide a day, called a *diurnal tide*, which has a period of nearly 25 hours.

The Sun also raises tides, which are only half the magnitude of lunar tides and have periods of 12 and 24 hours. The overall tidal amplitudes depend on the relationship between the solar tide and the lunar tide. The tidal amplitude is at its maximum twice a month during a new and full moon, and at its minimum during the first and third quarters of the moon. Certain tidal basins can resonate with the incoming and outgoing tides, making the high tides higher and the low tides lower than they would be otherwise.

Most inhabitants of the tidal zone (FIG. 8-8) possess biological clocks set to the rhythm of the lunar day. The rhythms are characterized by repetitive behavioral or physiological events, such as feeding, that are synchronized with the tides. Since there are generally two tides for each lunar day, which is about 25 hours long, the rhythms are called *bimodal lunar-day rhythms*, in contrast to the unimodal solar-day rhythms of organisms attuned to the 24-hour solar day.

These biological clocks are important for survival because they give advance warning of the regular changes in certain aspects of the environment, such as nightfall or the return of the tides. Even under constant laboratory conditions without the effects of diurnal or tidal cycles, the clocks continue to function and the biological rhythms persist for some time.

Apparently, the tidal rhythms are not learned or otherwise impressed on the organisms by the tides themselves. Crabs raised from eggs in the laboratory and exposed only to diurnal conditions exhibit a distinct tidal component in their activity after their body temperatures are lowered. Also, crabs taken from areas not subject to tides and moved to a tidal flat quickly establish a tidal rhythm. It appears that the clock that measures the tidal frequency is innate, and only needs to be activated by some outside stimulus.

The capacity for rhythmic behavior in organisms is also an expression of the genetic code. Heredity

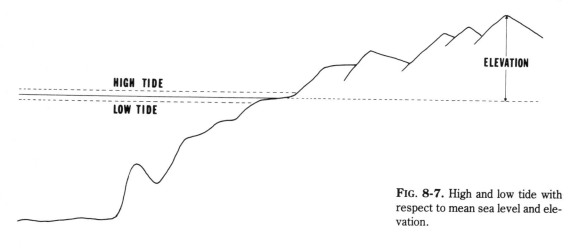

FIG. 8-7. High and low tide with respect to mean sea level and elevation.

FIG. 8-8. Intertidal exposure of chaotic blocks of sandstone.

determines whether an animal will be active during high or low tide. This does not mean that the environment does not play an important role in the establishment of a tidal rhythm, however. It is the schedule of the tides that determines the setting of the biological clock; therefore, animals transported to a different beach on a different ocean will synchronize their clocks to the new tidal conditions. Also, inhabitants living on beaches exposed to the open sea have their activity patterns shaped by the pounding surf.

Intertidal organisms living in protected bays are not as exposed to the vicissitudes of the sea and are controlled by more subtle conditions, such as the drop in temperature or pressure changes brought on by the incoming tides. Even in the absence of outside stimuli, an organism's clock continues to run accurately, but no longer controls its activity, and the animal operates independently from tidal influences until it returns to the sea and the clock takes over again. Like all clocks, the accuracy of biological clocks is not altered by changes in temperature or chemistry of the environment, nor do they just pertain to intertidal organisms but also the entire kingdom of life.

THE ORBITAL CYCLES

The amount of solar energy impinging on the Earth's surface is governed by the shape of the Earth's orbit around the Sun, the tilt of its axis, and the precession of its axis, and is called the *Milankovitch model of orbital variations* (FIG. 8-9) for the Yugoslav astronomer Milutin Milankovitch, who proposed the theory in 1941. These orbital variations change the climate by altering the amount of solar energy the Earth receives at different latitudes and seasons. For instance, currently the Earth is further from the Sun during summer in the Northern Hemisphere and closest during winter. The difference between aphelion and perihelion is about 3 million miles, which translates into about a 7 percent difference in the amount of sunlight the Earth receives between the two seasons. Therefore, if there were no other factors involved, such as the effect of ocean currents, the winters in the Southern Hemisphere would be colder than the northern winters.

If the Earth had a circular orbit, it would maintain a constant distance from the Sun of 93 million miles, and the total amount of solar energy the Earth received would be the same in all seasons. On the other hand, if the Earth's orbit were extremely elliptical, the difference in solar input could be as much as 30 percent. The southern winter would be much more severe than the northern winter, and the northern summer would also be much cooler. A complete orbital cycle from circular to elliptical and back again takes roughly 100,000 years.

The Earth's axis of rotation is inclined to the plane of its orbit around the Sun, called the *ecliptic*, at an angle of 23.5 degrees. As the Earth swings around the Sun, the position of the Sun relative to the equator changes, giving the Earth its seasons (FIG. 8-10). Any change in the axial tilt angle will shift the position where the Sun is overhead during the seasons. The lesser the degree of axis tilt, the more confined the Sun is to the tropics and the lesser the difference among seasons. Conversely, the greater the tilt angle, the greater the Sun's travel between the equator and the poles and the greater the difference among seasons.

The incline of the Earth's axis has varied from 22 to 24.5 degrees, and the effect is as though the Earth were nodding up and down, which is why the process is called *nutation*. The Earth completes one full nutation cycle every 41,000 years. For the past 10,000 years, the degree of tilt has been lessening, which could produce cooler summers and milder winters.

The axis of rotation also *precesses*, or wobbles, like that of a toy top, and is called the precession of the equinoxes. A line drawn through the Earth's axis describes a circle in the heavens that rotates clockwise, or in the opposite direction of the Earth's rotation. The Earth completes one precession cycle every 26,000 years. When combined with the rotation of the ellipse of the Earth around the Sun, the actual period is closer to 22,000 years.

Presently, Earth's axis points in the direction of Polaris, the North Star. Toward the end of the last ice age, the axis of rotation pointed to Vega, the brightest star in the constellation of Lyra. Thus, Earth's axis of rotation was 47 degrees in the opposite direction from where it is now and the seasons were completely reversed, with the Northern Hemisphere experiencing winter May through August. Primitive man living in the Northern Hemisphere would have then been able to see those constellations that are presently seen only in the Southern Hemisphere. Should man still be around 10,000 years from now, Christmas might occur during summer in the Northern Hemisphere.

Variations in the orbital motions combine to produce the overall changes in the pattern of solar radi-

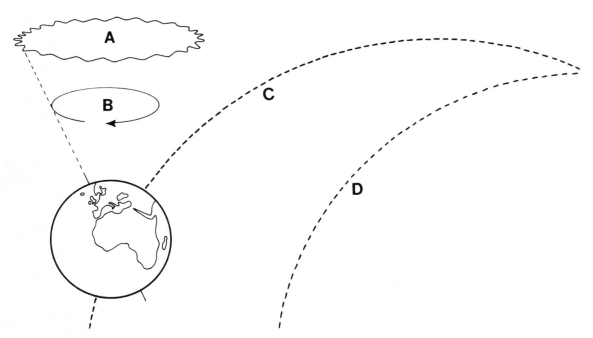

FIG. **8-9.** The Milankovitch model of orbital variations. A = nutation, B = precision, C = elliptical orbit, D = circular or bit.

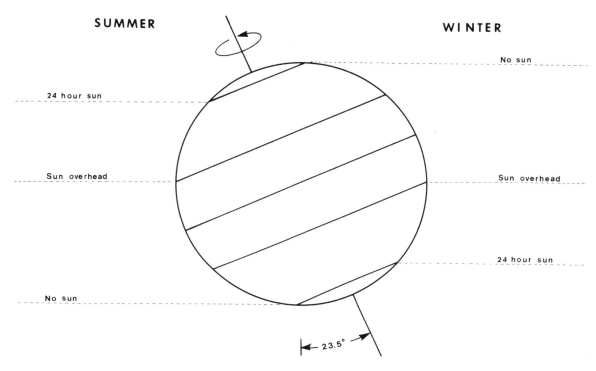

SUMMER   WINTER

No sun

24 hour sun

Sun overhead   Sun overhead

24 hour sun

No sun

23.5°

FIG. 8-10. The effect of the Earth's ecliptic on the seasons.

ation falling on the Earth. Furthermore, small variations in solar insolation could cause large changes in climate. If the Northern Hemisphere happened to be in summer when a highly elliptical orbit took it further away from the Sun and the tilt of the axis was at a minimum, the summers could be cool enough for the previous winter's snows to fail to melt. When winter returns, it would be milder but stormier, and more accumulations of snow would be added, until after several seasons, continental-size glaciers would start to form. Once established, the ice sheets would grow slowly for a long period, and then as the orbital motions favored a warmer northern climate, the ice sheets would melt rapidly.

The ice ages appear to have come and gone about ten times during the last million years, or about one every 100,000 years, which also happens to be the same period as Earth's orbital cycle. In addition, small fluctuations superimposed on the principal cycle might correspond to influences by the other orbital motions, which could reinforce or subtract from

the climatic effects of the 100,000-year cycle. Once the glaciers are in place, they are hard to get rid of because of their large *albedo effect*. That is, they reflect sunlight back into space that would otherwise go to warm the Earth.

THE CARBON CYCLE

The most important cycle with regard to life on Earth is generated by the Earth itself and is sometimes referred to as the *rock cycle*. Presently, carbon dioxide makes up about 0.035 percent of the atmosphere, which amounts to about 700 billion tons of carbon. It plays a critical role as a primary source of carbon, which is fixed by photosynthesis in green plants and therefore provides the basis for all life. Carbon dioxide traps heat that would otherwise escape into space and acts somewhat like a thermal blanket. In this respect, it plays an important role in regulating the temperature of the Earth. If more carbon dioxide were taken out of the atmosphere than is replenished, the Earth would cool down. If more carbon dioxide were generated than taken out of the atmosphere, the Earth would heat up. Therefore, any changes in the carbon cycle could have profound effects on the climate.

The amount of carbon dioxide in the early atmosphere was thousands of times greater than it is today because at that time, the sunlight was too feeble to maintain the temperature of the Earth within tolerable limits for the early organisms. During the steamy Cretaceous period when the dinosaurs roamed practically from pole to pole, the amount of carbon dioxide in the atmosphere was much greater than it is today. An atmosphere rich in carbon dioxide also might have been responsible for the luxurious plant growth during that period.

The *biota*, which is all living things on the surface of the Earth, and *humus*, which is dead organic matter in the soil, hold several times more carbon than is in the entire atmosphere. The harvest of forests, the extension of agriculture, and the destruction of wetlands not only destroy wildlife habitat but also speed the decay of humus, which is converted into carbon dioxide that enters the atmosphere. Also, agricultural lands do not store as much carbon as the forests they replace.

Forests are more extensive and conduct more photosynthesis than any other form of land vegetation. They incorporate from ten to twenty times more carbon per unit area than cropland or pastureland. Forests also have the potential to store carbon in quantities large enough to significantly affect the amount of carbon dioxide in the atmosphere.

The forests are being destroyed at an alarming rate to make way for agriculture in order to feed a hungry world. Half the forests in the civilized world are already gone, and every year an area of forest about the size of Ohio disappears. As the stores of carbon in the trees are being released into the atmosphere, the concurrent reduction of the forests is weakening their ability to remove excess carbon dioxide from the atmosphere.

The oceans contain more than 50 times as much carbon dioxide as the atmosphere, mostly as dissolved bicarbonate. Atmospheric carbon dioxide is continuously being dissolved in the oceans by surface-wave action, and the concentration of carbon dioxide in the upper 250 feet of the ocean is as much as there is in the entire atmosphere. Microorganisms in the sea use this rich store of carbon dioxide to build their skeletons, which are composed of calcium carbonate. When the microorganisms die, their shells are either reabsorbed by the seawater

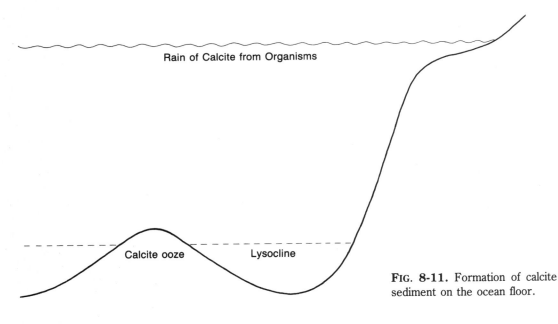

Rain of Calcite from Organisms

Calcite ooze   Lysocline

FIG. 8-11. Formation of calcite sediment on the ocean floor.

or accumulate on the ocean floor and form a calcite ooze (FIG. 8-11). This ooze becomes buried under thick sediments and solidifies into layers of lime-stone, which permanently locks up the carbon in the oceanic crust.

The abyssal region, by virtue of its great volume, holds the vast majority of the free carbon dioxide available in the world, and the capacity of this region of the ocean to store carbon dioxide is almost limitless. Because carbon dioxide moves from the atmosphere into the ocean at a slow, steady rate, rapid increases in atmospheric carbon dioxide such as during the current industrial age cannot be dealt with quickly enough, and the level of carbon dioxide in the atmosphere rises.

Carbonate rocks like limestone, which were formed by the inundation of the sea onto the continents, became a permanent repository of carbon that expanded as the continents grew. Exposures of limestone on the surface are weathered by rain containing a weak carbonic acid from the reaction of rainwater with carbon dioxide. The dissolved calcium carbonate is then returned to the sea by way of streams and rivers, completing the final leg in the hydrologic, or water, cycle.

Carbonate rocks in the oceanic crust are consumed in the subduction zones, where they descend into the mantle. The extreme temperatures in the mantle heat the carbonate rocks and drive off the carbon dioxide, which then becomes an important volatile in magma. The magma works its way to the surface and is erupted by volcanoes or flows out as

(Photo by Jim Vallance, courtesy of USGS)

FIG. 8-12. Mushroom cloud of ash rising from Mount St. Helens on July 22, 1980.

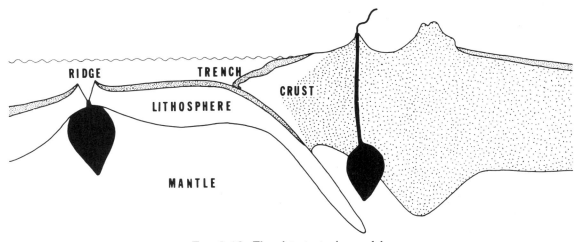

FIG. 8-13. The plate tectonics model.

Table 8-1. Major Changes in Sea Level.

DATE	SEA LEVEL	HISTORICAL EVENT
2200 B.C.	Low	
1600 B.C.	High	Coastal forest in Britain inundated by the sea.
1400 B.C.	Low	
1200 B.C.	High	Egyptian ruler Ramses II builds first Suez canal.
500 B.C.	Low	Many Greek and Phoenician ports built around this time are now under water.
200 B.C.	Normal	
A.D. 100	High	Port constructed well inland of present-day Haifa, Israel.
A.D. 200	Normal	
A.D. 400	High	
A.D. 600	Low	Port of Ravenna, Italy, becomes land-locked. Venice is built and is presently being inundated by the Adriatic Sea.
A.D. 800	High	
A.D. 1200	Low	Europeans exploit low-lying salt marshes.
A.D. 1400	High	Extensive flooding in low countries along the North Sea. The Dutch begin building dikes.

new oceanic crust at spreading centers. The carbon dioxide, along with water and other gases that are no longer under pressure, are released from the magma sometimes explosively (FIG. 8-12) and returned to the atmosphere or the ocean. Since nowhere is the oceanic crust more than 5 percent of the Earth's age, vast quantities of rocks are circulated through the mantle in a gigantic recycling plant (FIG. 8-13), which processes all the oceanic crust every 160 million years. Therefore, the Earth's ability to recycle carbon is perhaps the most fundamental requirement for maintaining the balance of nature.

9

Ice on the World

ICE ages have affected life on Earth almost from the very beginning, and it is even argued that life itself might have changed the climate sufficiently to bring on the ice. When the first microscopic plants developed, they replaced the carbon dioxide in the atmosphere with oxygen. The loss of this important greenhouse gas caused the climate to cool, which brought on the first known ice age about 2 billion years ago.

Another glacial episode, which occurred some 260 million years ago, might have been triggered by the spread of forests on the land, as adaptations allowed plants to live and reproduce out of the ocean. The Earth cooled as the forests removed atmospheric carbon dioxide and converted the carbon into organic matter, which was then buried in the sediments. This burial of carbon dioxide in the geologic column might have been the key to the onset of two other glacial epochs: one at about 700 million years ago, which was perhaps the greatest ice age of them all, and the most recent one, which began about 2 million years ago.

The positions of the continents influenced the ice ages as well, and a large amount of land near the poles might have allowed a buildup of glacial snow and ice. Global tectonics might have triggered the ice ages by volcanic activity or seafloor spreading, which drew oxygen out of the oceans and atmosphere so that more organic carbon was preserved in the sediments and not returned to the atmosphere by living organisms.

THE COMING OF ICE

It is hard to conceive of a world nearly half covered with ice, yet geological evidence indicates that major glaciers have traveled over much of the land surface at least four times during the Earth's life. Most of this evidence comes from deposits of glacial rocks, called *moraines* and *tillites*, (FIG. 9-1), which have been discovered and dated at various times throughout many parts of the world. Analysis of deep-sea sediments and glacial cores have provided accurate information about some of the events taking place during the most recent ice age.

In order for the ice sheets to form, the climate

115

was cooler, but not necessarily a lot cooler than it is today. During the last ice age, the average daytime temperatures were probably not more than ten degrees Fahrenheit lower than they are today. Once the ice was established, it became self-perpetuating and seemed to be able to control the climate to maintain its own existence.

One of the feedback mechanisms by which ice sustained itself is through the albedo effect, that quality of an object that enables it to reflect sunlight. The albedo of an object is dependent on its color. Because snow is white, it has a high reflectance, as anyone who has skied in the bright sunshine knows. Most of the sunlight is reflected out to space and very little goes to heat the snowpack. However, extremely cold temperatures also inhibit the precipitation of snow, which is why although Antarctica is covered with up to 2 miles of ice, it is also one of the dryest as well as the coldest spots on Earth.

There are a number of triggering mechanisms that could initiate the coming of ice. Any factor that reduces the amount of solar energy impinging upon the Earth and lowers the Earth's temperature is a good candidate. The reduction in the amount of sunlight might come from the Sun itself. Although the amount of sunlight now appears to be quite stable, a slightly lowered output could have been responsible for events like the Little Ice Age when advancing glaciers chased people out of the northlands of Europe two to three centuries ago. Disregarding the Sun as the villain, the orbital variations of the Earth, discussed in Chapter 8, could affect the amount of sunlight the Earth receives from one season to the next, but they do not actually reduce the average amount of solar input during the entire year. The solar system in its orbit around the galaxy has encountered enormous clouds of dust in interstellar space at least 100 times during its lifetime. Such dust falling into the Sun could disturb its normal activity, and it could also affect the Earth's climate by causing ice clouds to form in the upper atmosphere which would prevent some of the Sun's heat from reaching the Earth's surface.

The culprit might even be the Earth. The rise of mountains as a result of the collision of crustal plates could push large portions of the land above the snow line and form glaciers, which could travel down to the lowlands. A large number of volcanic eruptions could eject huge quantities of ash and dust into the atmosphere and effectively block out the

(Photo by I.C. Russell, courtesy of USGS)

FIG. 9-1. Terminal moraine at the margin of a glacier, Deschutes County, Oregon.

Sun's rays, which could bring on the ice sheets. It has even been suggested that the weight of the ice on the continents forces magma out of its chambers and onto the surface, like toothpaste squeezed from a tube.

One of the hazards of increasing carbon dioxide content in the atmosphere through the combustion of fossil fuels and the destruction of the forests is the creation of what is called a *glacial surge*. As the present climate continues to warm, the West Antarctic ice sheet could become highly unstable. Friction at the base of the ice sheet would be lost because of subglacial lakes and streams, which act as a lubricant, and large parts of the ice sheet could be sent crashing into the ocean. Also, the melting of glacial ice in both polar regions could raise the sea level high enough so that West Antarctic ice shelves (FIG. 9-2), pinned to undersea islands, would be ripped from their moorings and drift out to sea. This occurrence would further raise the sea level, setting loose more ice.

As the sea level continued to rise, the shoreline would advance onto the continents with disastrous consequences to most major cities of the world. The increased area of ice in the Southern Ocean could form a gigantic ice shelf, covering as much as 10 billion square miles. This ice shelf would greatly increase the Earth's albedo, which would diminish temperatures in the Southern Hemisphere. With enough ice in the ocean, seawater could be cooled dramatically, and ocean currents and weather patterns could be affected sufficiently to bring on a new ice age.

No matter which way the glaciers began, their effects on life would be enormous. The living space of warmth-loving species would be drastically reduced to areas narrowly confined to the tropics. This shrinkage of viable habitat is why major extinctions usually followed such an event, and those species unable to migrate or adapt to the colder conditions were generally the losers.

Because lowered temperatures slow down the rate of chemical reactions, biological activity during a major glacial event would be expected to function at a lower energy state. The contradiction to this would be the development of new species that be-

FIG. 9-2. The frozen continent of Antarctica. Stippled areas are ice shelves.

came better adapted to the cold, and it is even speculated that our own ancestors, which developed during the last ice epoch, were one of these new species.

The dinosaurs apparently became extinct because the climate at the end of the Cretaceous period was too cold for them and their favorite food to exist. Mammals, which carry around their own sources of heat, were particularly well adapted to the colder conditions, and some living aquatic mammals even required the cold in order to survive. Other animals, such as birds and aquatic species, developed migratory habits, which allowed them to escape the frigid weather during the winter for better living conditions in the opposite hemisphere.

PRECAMBRIAN ICE

The *Gaia Hypothesis*, named after the Greek goddess of the earth, was introduced by the British chemist James Lovelock in 1979 and contends that life can control its own destiny, to some extent, by exerting a certain influence on the climate (FIG. 9-3). First, it is necessary to define *life*. In physical terms, life is a huge, intricate molecular machine that seems

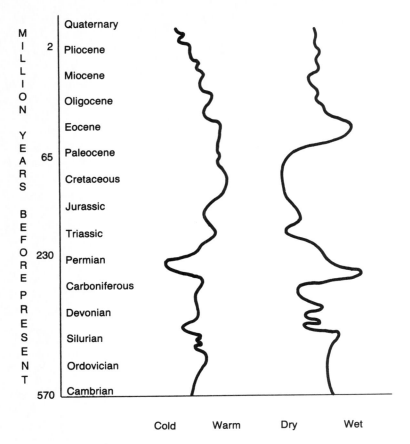

| | | | Cold | Warm | Dry | Wet |

FIG. 9-3. The climate through time.

to overcome, at least for a little while, the Second Law of Thermodynamics, which states essentially that every form of order dissolves into chaos.

Life manages to go against the flow of a universe that is steadily running downhill, but the uphill struggle comes at the expense of a great deal of energy, which it obtains from the Sun. This energy is manifested by the presence of large amounts of oxygen in the atmosphere since without life, chemical reactions would have run downhill and the Earth's oxygen would long since have become bound to other elements, such as iron.

Life appears to maintain oxygen and carbon dioxide in a perfect balance, and too much of one with respect to the other could have disastrous consequences. It almost seems as though life were conscientiously controlling its own environment for optimum living conditions. Actually, however, it is more like an unthinking regulator similar to a home thermostat, which keeps the house at a comfortable temperature no matter how harsh the weather is outside.

Photosynthesis probably got underway as early as 3.5 billion years ago, but any oxygen that was produced by photosynthetic organisms was quickly snatched up by chemical reactions that permanently locked it up in the crust. It was not until about 2 billion years ago that these oxygen sinks held as much oxygen as they possibly could, and the gas began to slowly build up in the ocean and atmosphere.

In addition to the generation of oxygen, simple plants removed carbon dioxide from the environment and buried it permanently in the geologic column. It was also about this time that plate tectonics began to operate extensively, and carbonaceous sediments along with the oceanic crust were thrust deep within the Earth's interior. The growing continents also stored large quantities of carbon dioxide in thick deposits of carbonaceous rocks. The elimination of this important greenhouse gas, which was responsible for maintaining the Earth's warm temperature, caused the Earth to cool substantially. The continental landmass, which was much smaller than it is today, probably was located near one of the poles where huge ice sheets grew.

Although this was the first ice age the world had experienced, it was not the most extensive. The greatest glaciation the planet has ever endured took place about 700 million years ago, when nearly half the world was encased in ice. The climate was so cold that ice sheets and *permafrost*, permanently frozen ground, lay near equatorial latitudes. At this time, no plants grew on the barren land surface, and only simple unicellular plants and animals existed in the sea. Thick sequences of Precambrian tillites, which are an admixture of boulders and pebbles with a clay matrix consolidated into solid rock deposited by glacier ice, are known to exist on every continent (FIG. 9-4). In the Lake Superior region of North America, tillites have been found to be 600 feet thick in places and range from east to west for a distance of at least 1,000 miles. In northern Utah, tillites mount up to a thickness of 12,000 feet, giving the impression that there were a series of ice ages following one another closely. Similar tillites were found among Precambrian rocks in Norway, Greenland, China, India, southwest Africa, and Australia.

With the end of the glaciation and the retreat of the ice, life began to proliferate in the ocean on a grand scale. As the oceans warmed, life processes speeded up and there was an explosion of new species. The first multicellular animals appeared, and for the first time, fossils began to appear in relative abundance.

PALEOZOIC ICE

Another substantial carbon dioxide sink was the great coal forests of the Upper Paleozoic era. Plants

FIG. 9-4. Location of Late Precambrian glacial deposits.

invaded the land and rapidly spread to all parts of the Earth by about 400 million years ago. Lush forests, which grew during the Carboniferous period (FIG. 9-5), absorbed large quantities of carbon dioxide, which they used in their growth. Rapid burial under anaerobic conditions converted the carbon in the vegetation into thick seams of coal, and like carbonaceous rocks, these seams also buried carbon dioxide in the geologic column until the coal was later dug up and used by man. Presently, the coal is being burned in massive quantities in coal-fired plants throughout the world, releasing stored solar energy and returning carbon dioxide to the atmosphere.

The glaciations of the Late Ordovician period, 440 million years ago, and the Permo-Carboniferous, 290 million years ago, might have been influenced by a reduction of atmospheric carbon dioxide, equal to about a quarter of what it is today. The notion that glacial epochs could be a result of changes in carbon dioxide is not a new idea, but only recently have scientists acquired enough information on global geochemical cycles to determine what might have caused the change in the concentration of carbon dioxide in the atmosphere. New data from deep-sea cores indicated that variations in the amount of carbon dioxide preceded changes in the extent of the more recent ice sheets, and it is not unexpected that the earlier glacial epochs were similar. The variations of carbon dioxide levels might not be the sole cause of glaciation, but when combined with other processes, they could have been a strong influence.

The Late Paleozoic era was also a period of extensive mountain building, which raised a lot of ground to higher elevations where glaciers could be nurtured (FIG. 9-6). Glaciers might have formed and

FIG. 9-5. Approximate position of the equator during the Carboniferous period.

C-1. The Sonora Desert, Arizona, in bloom. (USGS)

C-2. Lizards—living relatives of the extinct dinosaurs. (NPS)

C-3. Buffalo roaming by Old Faithful at Yellowstone National Park.

C-4. Fossil of Oreodon with unborn twins. (RPI)

C-5. Technicians reliefing leg and shoulder bones of Apatosaurus. (NPS)

C-6. Ancient pictograph of a bighorn sheep near Jones Creek, Utah. (NPS)

C-7. Glacial valley, Glacier National Park, Montana. (USGS)

C-8. Meteor Crater in Arizona. (USGS)

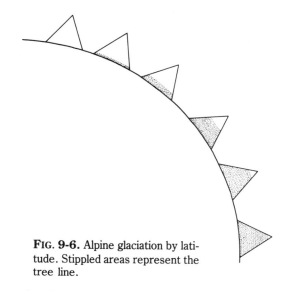

FIG. 9-6. Alpine glaciation by latitude. Stippled areas represent the tree line.

persisted on continents even at low latitudes as long as the elevation was high enough.

The supercontinents of Gondwanaland and Laurasia converged into the crescent-shaped continent of Pangaea, whose landmass extended almost from pole to pole. The continental collisions crumpled the crust and pushed up huge masses of rocks into several mountain chains throughout many parts of the world. In addition to the folded mountain belts volcanoes were prevalent, and unusually long periods of volcanic activity could have blocked out the Sun with clouds of volcanic dust and aerosols and thereby lowered surface temperatures.

As the continents rose higher, the ocean basins dropped lower. The change in the shape of the ocean basins greatly affected the course of ocean currents, which in turn had a profound effect on the climate. The continental margins became less extensive and narrower, confining marine habitat to near-shore areas, which might have influenced the great extinction at the end of the era. The land, which was once covered by great coal swamps, became drier and the climate grew colder.

During the breakup of Pangaea, when Gondwanaland separated from Laurasia and the southern landmass passed into the south polar regions, glacial centers expanded in all directions. Land existing near the poles is often the cause of extended periods of glaciation because land located at higher latitudes usually has a higher albedo and a lower heat capacity than the surrounding oceans and encourages the accumulation of ice.

Ice sheets covered large portions of east central South America, South Africa, India, Australia, and Antarctica (FIG. 9-7). Early in the ice age, the maximum glacial effects were in South America and South Africa. Later, the chief glacial centers moved to Australia and Antarctica, providing good evidence for the existence of Gondwanaland and that it wandered en masse around the South Pole. In Australia, marine sediments were found interbedded with glacial deposits, and tillites were separated by seams of coal, which indicated that periods of glaciation were interspersed with warm interglacial spells. In South Africa, the Karroo Series, composed of a sequence of Late Paleozoic tillites and coal beds, reached a total thickness of 20,000 feet. Among the coal, which is the best in Africa, are fossil leaves of glossopteris (FIG. 9-8), whose existence there and on the other southern continents provides some of the best evidence for continental drift.

PLEISTOCENE ICE

The Pleistocene epoch, which began about 2 million years ago, saw a progression of ice ages, each one followed by a short interglacial period similar to the one we are living in now. The last ice age began about 100,000 years ago, peaked about 18,000 years ago, and retreated about 10,000 years ago, indicating that the ice took much longer to build up than it did to disappear.

The most recent glaciation is perhaps the best studied of all ice ages, mainly because each succeeding ice age has a tendency to erase much of the evidence of the previous ones. It is for this reason that it was so difficult earlier to test the Milankovitch theory of orbital variations and their effects on glaciation. New indirect evidence was found in the form of coral terraces running up coral islands or atolls. During the last ice age, about 5 percent of the Earth's water was locked up in glacial ice, which resulted in the lowering of the sea level by as much as 300 feet (FIG. 9-9). Coral, which lives in the tropics near the surface of the sea, fluctuates in

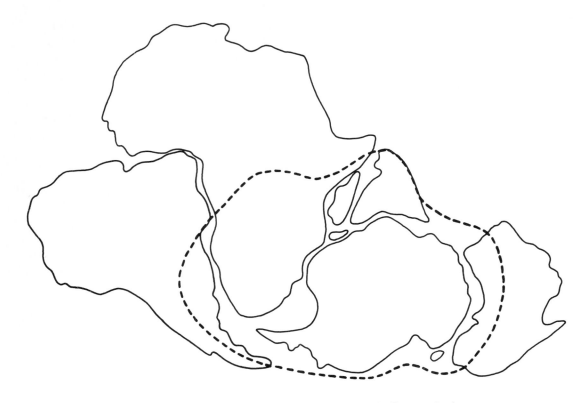

FIG. **9-7.** Extent of Late Paleozoic glaciation in Gonwanaland.

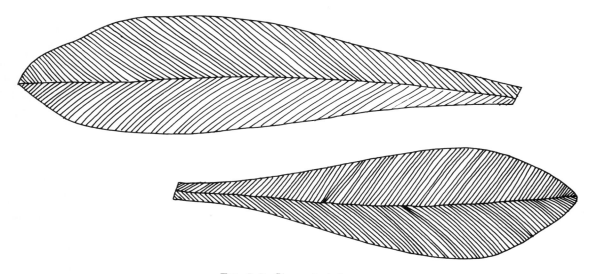

FIG. **9-8.** Glossopteris leaves.

height according to the sea level. The lowering of the sea caused erosion of the coral reef down to the new sea level, and the rising of the sea again when the glaciers melted caused new coral to grow on top of older coral, forming a terrace. Alternating sea level changes corresponding to the waxing and waning of the glaciers produced a staircase structure of coral growth that matched the ages of the glacial periods.

Further proof for rapid and rhythmic successions of glaciation came from analyzing fossil shells from core samples of the ocean floor for their heavy oxygen content. The ratio of heavy oxygen-18 to oxygen-16 is an indication of the ocean's past temperature and provides a means of accurately dating the ice ages. According to these data, there have been about ten ice ages in the past million years, thus providing concrete evidence for the 100,000-year Milankovitch orbital cycle.

The most persuasive argument that large parts of the northern continents were once covered with

ice (FIG. 9-10) came from the Swiss geologist Louis Agassiz early in the nineteenth century. In a large U-shaped valley in the Swiss Alps, Agassiz discovered rocks that had been scratched and polished by the passage of ice (FIG. 9-11). He identified heaped deposits of sand and gravel, called *moraines*, and huge boulders, called *irratics* (FIG. 9-12), that were transported over great distances. Agassiz envisioned that the valley had been filled with a glacier 1 mile thick. As the glaciers descended from the mountains and spread across most of Europe, they killed everything in their path. Most scientists of his day ridiculed Agassiz's ideas and claimed that the rocks were simply carried there by the Great Flood. However, the morphology of the deposits was not that created by floodwaters, but resulted from retreating glaciers, which dropped their sediment load at the point where the ice melted.

In 1846, Agassiz immigrated to America and found more impressive evidence of intense glaciation. Virtually all of the continent north of the Ohio

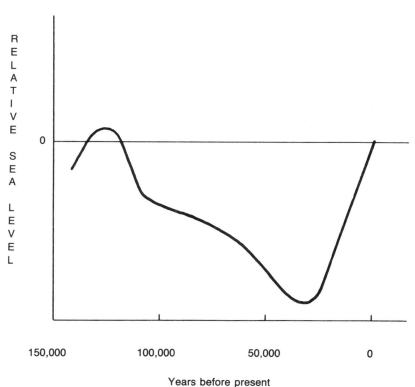

FIG. 9-9. Sea level through time.

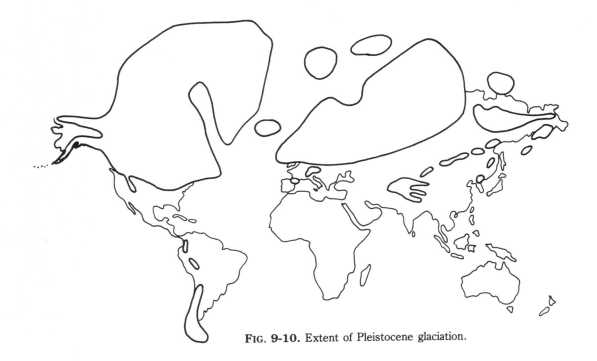

FIG. 9-10. Extent of Pleistocene glaciation.

FIG. 9-11. Glaciated valley, Glacier National Park, Montana.

(Photo by George A. Grant, courtesy of National Park Service)

FIG. 9-12. Erratic glacial boulder in the Sierra Nevadas, Fresno County, California.

and Missouri rivers had been glaciated. At the height of the ice age, the North American ice sheet covered 5 million square miles up to 2 miles thick or more.

There were two main glacial centers in North America. The largest ice sheet spread out from Hudson Bay and reached north to the shores of the Arctic Ocean and south to bury all of eastern Canada, New England, and much of the rest of the northern half of the midwestern United States. A smaller ice sheet originated in the Canadian Rockies and engulfed western Canada, parts of Alaska (FIG. 9-13), and small portions of the northwestern United States. The ice buried the mountains of Wyoming, Colorado, and California, and rivers of ice connected them with mountains in Mexico.

There also were two major glacial centers in Europe. The largest ice sheet radiated from northern Scandinavia and covered most of Great Britain and large parts of northern Germany, Poland, and European Russia. A smaller ice sheet centered in the Swiss Alps covered parts of Austria, Italy, France, and southern Germany. In Asia, the ice sheets occupied the Himalayas and parts of Siberia.

In the Southern Hemisphere, only Antarctica had a major ice sheet, which had no place to go except into the ocean. Small ice sheets expanded in the mountains of Australia, New Zealand, and the Andies of South America. Just about everywhere, alpine glaciers existed on mountains that are presently ice free.

The temperature at the Earth's surface averaged over the globe and over the seasons was about 10 degrees Fahrenheit lower than today's average. The cold weather and advancing ice forced animals and humans to migrate to southern lands. Ahead of the slowly advancing ice sheets, which averaged a few hundred feet per year, lush deciduous forests gave way to evergreen forests, which themselves gave way to barren tundra.

The lowered temperatures also caused less water to evaporate from the oceans and reduced the average amount of precipitation on the land. Because very little melting took place in the cooler summers, only a small amount of snowfall was needed to sustain the glaciers. The lower precipitation levels also increased the spread of deserts in many parts of the world, and desert winds were more blustery than they are today. Large numbers of icebergs calved off glaciers entering the sea, and like ice cubes in a cold drink, they kept the ocean's surface temperature cool. In addition, the bottom of the ocean had been steadily cooling down since the Cretaceous so that today, it is near freezing (FIG. 9-14).

The lowered sea level extended the shoreline of the continents (FIG. 9-15) and brought several land bridges to the surface, which greatly aided the migration of animals and humans into various parts of the world. Adaptations to the cold climate allowed certain species of mammals to thrive in the northern lands that were not covered by glaciers. Giant mammals like the mammoth and mastodon roamed many parts of the world (FIG. 9-16). The reasons for the giantism might be similar to the reason many dinosaurs were so large and possibly included an abundant food supply and lack of predation.

As the glaciers began to retreat, about 16,000 years ago, there was a readjustment in the global environment as the cool but equable climate of the ice age gave way to the warmer but seasonal climate of today. This rapid environmental switch from glacial to interglacial caused a shrinking of the forests and an expansion of the grasslands. It might have disrupted the food chain of several of the large mammals, and deprived of their resources, they simply disappeared. It is often suggested that man, who was by this time an efficient hunter and roamed northward in response to the retreating glaciers, overhunted the slow, lumbering creatures similar to the way American expansion into the West decimated the bison almost to its extinction.

THE HOLOCENE INTERGLACIAL

One of the most dramatic climatic changes in the history of the Earth has occurred during the present interglacial time, known as the *Holocene epoch*, which began about 10,000 years ago. The collapse of the last ice age and the subsequent warm climate has left many puzzles, such as hippopotamus bones found in African deserts. Some of these baffling questions about the Holocene climate can be answered by analyzing various climate indicators, such as pollen grains recovered from ancient bogs and lake-bed sediments. These analytical methods convert observed abundances of plant and animal remains into measurements of the past climate by comparing them to present-day relationships be-

(Photo by Norman Herkenham, courtesy of National Park Service)

FIG. 9-13. Looking north up Yentna Glacier, Mount McKinley National Park, Alaska.

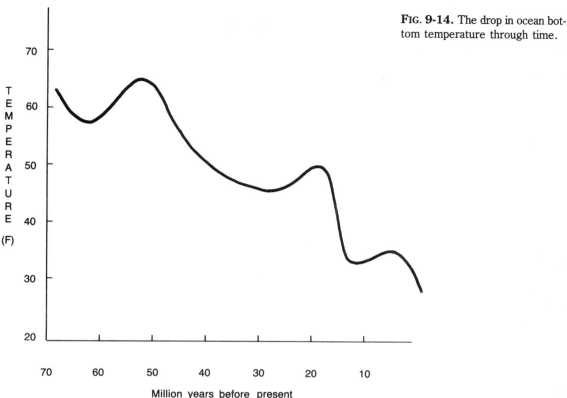

FIG. **9-14.** The drop in ocean bottom temperature through time.

Million years before present

tween species distribution and climate. For example, a combination of spruce and sedge fossil pollens is an indicator of a cold, dry climate, while fossil pollen of leafy herbs typical of the prairie of the Midwest suggests a warm, dry climate.

Another important continental climate indicator is lake-level fluctuations. Lakes act like natural rain gauges, and ancient rainfall amounts are implied by studying past shorelines and depth indicators such as the mineral, floral, and faunal composition of the lake sediments.

After some 90,000 years of gradual accumulation of snow and ice up to 2 miles thick in parts of North America and Europe, it all wasted away to nearly nothing in just a few thousand years. The rapid deglaciation might have been driven largely by forces other than simply the warming of the climate. The ultimate cause of the ending of the last ice age might have been changes in the Earth's orbit and

axial tilt. Nine thousand years ago, the Earth's axis of rotation was more tilted than it is today; the Earth came closest to the Sun in late July, instead of early January; and the annual range of the distances between the Sun and the Earth was greater. The net effect was 7 percent more sunlight over much of the Northern Hemisphere during the summer and 7 percent less during the winter, making the contrast between the seasons greater.

At least one-third of the ice melted between 16,000 and 13,000 years ago when the average global temperature rose about 10 degrees Fahrenheit. This rise in temperature culminated in the extinction of microscopic organisms called *foraminifera* as a result of a torrent of meltwater and icebergs that spilled into the North Atlantic, forming a cold, freshwater lid on the ocean. Then, the temperature fell again to near ice age levels, and ice sheets appeared to pause in mid-stride between

FIG. 9-15. Extended shoreline during the height of the Pleistocene Ice Age.

13,000 and 10,000 years ago. After that, the warming remained, and a second episode of melting led to the present volume of ice about 6,000 years ago.

As the ice began to shrink, the tropical regions of Africa and Arabia began to dry out, culminating in broadly distributed arid regions from 14,000 to 12,500 years ago. During a wet period from 10,000

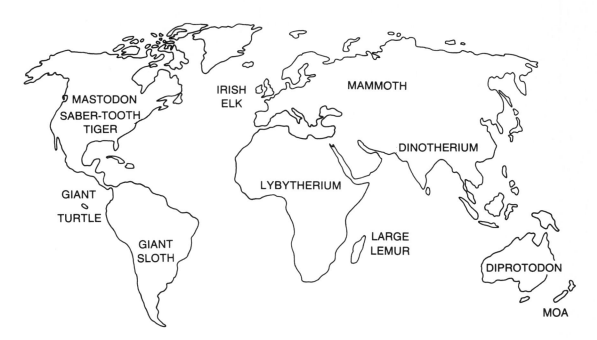

FIG. **9-16.** Giant animals of the Pleistocene.

———— Table 9-1. Chronology of the Major Ice Ages. ————	
TIME IN YEARS	**EVENT**
2 billion	First major ice age.
700 million	The great Precambrian ice age.
230 million	The great Permian ice age.
230–65 million	Interval of warm and relatively uniform climate.
65 million	Climate deteriorates, poles become much colder.
30 million	First major glacial episode in Antarctica.
15 million	Second major glacial episode in Antarctica.
4 million	Ice covers the Arctic Ocean.
2 million	First glacial episode in Northern Hemisphere.
1 million	First major interglacial.
100,000	Most recent glacial episode.
20,000–18,000	Last glacial maximum.
15,000–10,000	Melting of ice sheets.
10,000–present	Present interglacial.

to 5,000 years ago, some of today's African deserts became dotted with lakes. Lake Chad, which lies on the border of the Sahara desert, apparently swelled over ten times its present size. Swamps, long since vanished, once harbored hippopotamuses and crocodiles.

The long wet spell is believed to have been caused by the strengthening of the monsoons, which carry moisture-laden sea breezes inland over Africa, India, and Southeast Asia. The interior of continents 9,000 years ago warmed more in the summer, which strengthened the monsoon winds and increased rainfall.

In North America, the glaciers changed the atmospheric circulation and storm patterns over the continent and made it possible for the Mojave and nearby deserts of the southwestern United States to be wet enough to sustain woodlands even after the ice sheets began their retreat. From about 9,000 to 6,000 years ago, when the glaciers shrunk to a nearly insignificant chunk of ice in Labrador, precipitation over much of the Midwest dropped by as much as 25 percent, while the mean July temperature rose by as much as 4 degrees Fahrenheit. Also, it appears that the postglacial eastern and southeastern United States was no warmer 6,000 years ago than it is today.

The last interglacial period appears to have been warmer than the current one according to paleoclimate studies, which analyzed marine microfossils in cores of seafloor sediments for their levels of car-bon dioxide. It appears that there were higher atmospheric carbon dioxide concentrations during the preceding warm interglacial than the present one.

Another paleoclimate indicator is the size of sand grains that were deposited over the past 260,000 years in Lake Biwa, Japan, one of the oldest lakes on Earth. The rise and fall of the sediment size kept in step with the advance and retreat of the ice sheets. The grain sizes reflected erosion rates, which in turn depended on climate changes in rainfall, temperature, and wind speed. High precipitation rates increase erosion and therefore the amount of coarse sand grains carried to the lake. Ancient river beds, which carried alluvium to the sea, also were used as a source of information on changing climatic conditions of the past.

These and other tests indicated that glacial periods were many times longer than interglacial periods, which generally lasted only about 10,000 years. The fact that the present interglacial period has existed about 8,000 to 10,000 years means that it has just about run its course. Although the last interglacial might have been warmer than this one, the warm climate did not appear to stop the onset of the last ice age. Even the large amount of carbon dioxide being pumped into the atmosphere by industrialization and the extension of agriculture seems to be powerless in stopping the ice sheets, and ice ages will probably come and go just as they have done over the past 2 million years.

10

The Great Heat Engine

THE importance of water to life is so obvious that it is often overlooked. Since Earth has an abundance of water as well as a highly diverse and prolific biosphere, it appears that water must be a prime prerequisite for life. There is so much water that too often it is taken for granted, and valuable water resources become wasted or polluted. Each day, the Sun evaporates a trillion tons of seawater that condenses into rain and allows life to flourish on the land. To put this number into some sort of perspective, imagine a bucket of water 7 miles wide and 7 miles high pouring down every day. This is roughly 25 inches of rain annually when averaged over the entire surface of the globe.

Some places get more precipitation than others (FIG. 10-1). Rain forests receiving 200 inches or more per year, while deserts might receive only 2 inches per year. Most of this precipitation falls directly into the ocean, but an abundant amount of water lands on the continents, which hold enough freshwater to fill the Mediterranean Sea 10 times over.

Three percent of the Earth's water is fresh, but 75 percent of that is locked up in glacial ice. The remaining freshwater is in the atmosphere, rivers, lakes, aquifers, soil, and plant and animal tissues. As far as terrestrial life is concerned, these are the most important sources of water, for without them the Earth would become a dry, desolate wasteland similar to a martian landscape.

A DELICATE BALANCING ACT

The Earth receives only about one-billionth of the total output of the Sun. What it does with its meager portion of the Sun's energy can either make the planet exceedingly cold or excessively hot, and life could not exist here. If the Earth had no atmosphere to distribute the Sun's heat over the surface, daytime highs could be hot enough to melt lead and nighttime lows would be colder than Antarctica in the dead of winter. If average solar energy received on the ground in a year was spread evenly around the world, an area the size of a football field

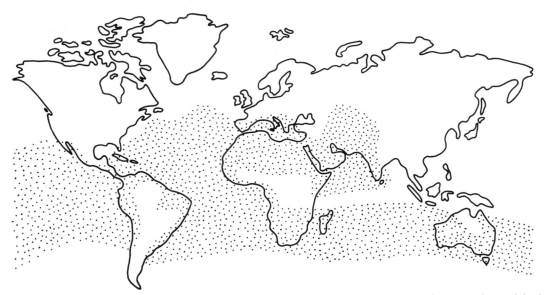

FIG. 10-1. The precipitation-evaporation balance of the Earth. In stippled areas, evaporation exceeds precipitation.

would receive over 1 million watts. On the other hand, the energy radiating from the Earth's interior is insignificant and would amount to only a couple hundred watts for the same area.

When sunlight strikes the Earth's surface, it is converted into infrared energy, which is reflected back out into space. The Earth must reradiate into space exactly the same amount of solar energy it receives, or it will become intolerably hot. If it reradiates too much solar energy, it will become intolerably cold. This delicate balancing act is called the *heat budget* and is responsible for maintaining the Earth's temperature within the narrow range that makes life possible.

The heat budget is mainly dependent on the Earth's albedo, or reflectance factor, and some objects reflect solar energy while others absorb it, depending on their color. Light-colored objects like clouds, snow fields, and deserts reflect more solar energy than they absorb, while dark-colored objects like oceans and forests absorb more solar energy than they reflect. About 90 percent of the solar energy absorbed by the oceans is used to evaporate seawater. Ultimately, this energy is given up and lost to space when the water vapor condenses into rain.

The angle of sunlight incident on the Earth's surface also determines how much solar energy is absorbed and how much is reflected. In the tropics, the Sun's rays strike the Earth at a high angle, which increases absorption of solar radiation. In the polar regions, the Sun's rays strike the surface at a low angle, and the solar radiation simply glances off into space. This angle of incidence as well as their high albedo keeps the polar regions covered in ice year round.

Solar energy is also scattered sideways as a result of dust particles and aerosols in the atmosphere, mainly from volcanic activity, duststorms, forest fires, pollution, and micrometeorite bombardment. These fine particles in the air are responsible in large part for making the sky blue because this color is at the high end of the solar spectrum and much of it is scattered by the atmosphere. If it were not for the dispersion of light by the atmosphere, the sky would be as black as night and the Sun would be nothing more than a large star up in the heavens. By the time the Sun sets, its rays, which are then at a low angle, must pass through so much atmosphere that only the reds are seen. As a result, we see fantastic sunsets.

About 40 percent of the Sun's energy is

reflected back into space before it ever has a chance to heat the Earth. Most of this lost energy is reflected off the tops of clouds. Clouds have a high reflectance, similar to that of snow and desert sand. The underside of clouds also reflect escaping infrared energy back to the ground, which is the reason nights are warmer under an overcast sky and cooler under a clear one. *Aerosols*, which are fine solid or liquid particles injected into the atmosphere by natural causes such as volcanoes or by man-made pollutants, also block the Sun's heat from reaching the ground, while allowing infrared heat from the ground to escape into space. Thus, they act like a one-way mirror.

About half the solar energy reaches the ground, where it is absorbed by the soil and plants that use the reds and blues for photosynthesis, but have no use for the greens. The greens are reflected away, so plants appear this color.

Eventually, all the sunlight that manages to reach the surface of the Earth is converted into infrared energy, which is radiated out to space. If it were not for the greenhouse effect, which prevents all the infrared from departing the Earth, the planet would indeed become a very cold place to live.

THE WEATHER MACHINE

The general structure of the atmosphere (FIG. 10-2) is the primary controlling factor in the distribution of life on this planet. The atmospheric flow patterns distribute heat from the tropics to the poles, and if it were not for the circulation of the atmosphere, the tropics would be too steamy and the polar regions would be too frigid for habitation. Most of this interaction involves the lower 12 miles of the atmosphere, called the *troposphere*. This 12-mile blanket of air is constantly in motion. Warm, moisture-laden air in the equatorial regions ascends to the top of the troposphere, called the *tropopause*, where it is blocked from rising higher, and it moves laterally toward the polar regions. There, the buoyant air gives up its heat, cools, descends, and returns to the tropics.

This explanation is an over simplification, of course, and the actual exchange of heat from the tropics to the poles is conducted by cellular structures called *Hadley cells*, named after the English meterorologist George Hadley who proposed the theory of heat convection in 1735. Because of the rotation of the Earth, flowing air masses are forced into spirals by the Coriolis effect, which deflects

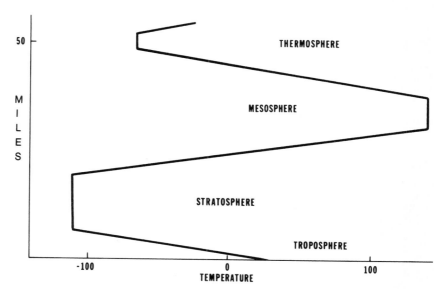

FIG. 10-2. Layering of the atmosphere.

them to the right in the Northern Hemisphere and to the left in the Southern Hemisphere. Low-pressure centers and hurricanes are vivid examples of these spiral structures (FIG. 10-3) and play an important role in distributing the Earth's heat. Unfortunately, they can also be very destructive. Every year they damage billions of dollars in property and take thousands of lives.

The clashing of great masses of air produces storms (FIG. 10-4). These weather systems are often born in and steered by the jet stream and surface topography; therefore, storm tracks frequently take well-traveled paths. There are cycles of meteorological events on time scales ranging from days to decades, and individual storms and days of extreme heat or cold come and go with periods of 3 to 10 days.

Tropical storms result when air from both sides of the equator rushes in, meets head on, and rises. As the warm, moist air gains altitude, it cools by expansion as the surrounding air becomes thinner and colder. The cooling causes the water vapor to condense into clouds, which releases thermal energy through the change of state of water. In turn, this

1600 11SE84 38A-1 02735 24643 MA32N79W-1

(Courtesy of NOAA)

FIG. 10-3. Hurricane Diana viewed from space.

FIG. 10-4. Global air currents are responsible for most of the weather.

energy causes the clouds to continue to rise. As cooling progresses, the clouds further condense into rain, releasing additional energy and causing the clouds to climb even higher. Tropical storms produce some of the heaviest rainfall in the world and are responsible for the equatorial forests and jungles of South America and Asia. They are also responsible for bringing life-giving monsoons to southern Asia and parts of Africa.

Similar collisions of air masses take place in the temperate zones, where polar fronts marching southward encounter warm air and force it to rise and condense into storm clouds. When warm currents that develop over the Gulf of Mexico or the Pacific Ocean flow northward, they form a *warm front*, and a collision with a mass of cold air forces the warm air to rise and produces stormy weather. Also, any long-term shift in the pattern of rising, low-pressure, or falling, high-pressure, air will determine whether there will be rain forests or dry deserts.

The weather over the United States has been getting cloudier, especially over the past several years. A comparison of the number of cloudless days in which an average of 10 percent or less of the daytime sky was obscured by clouds, fog, haze, or smoke was made for 45 major cities for the periods of 1900–1936 and 1950–1982, using data from the National Weather Service. In addition, an increase in cloudiness has been independently documented for the middle third of the United States. The results indicated that the second half of this century had more cloudy days than the first half. Some cities like Los Angeles, California, had an increase in cloudiness from 20 to 30 percent or more. Of all the cities checked, only Fort Worth, Texas, was sunnier, but the increase in sunshine was practically insignificant.

Several ideas have been put forward to explain the trend toward more cloudiness. One suggests that contrails from jet aircraft act as seeds to encourage condensation into cloud formations. Other data indicates that the polar weather front has shifted south, and this could instigate more clouds and storms. Pollution also could supply the microscopic particles for the condensation of cloud droplets. However, there seems to be a paradox here; for if pollution causes an increase in cloudiness, which produces cooler weather, while increased carbon dioxide causes a greenhouse effect, which makes the climate warmer, the two processes should cancel out each other.

THE WATER CYCLE

The average journey of a water molecule from the ocean through the atmosphere, across the land, and back into the ocean can take about 10 days. The round trip is much shorter in the tropical coastal zones, taking only a few hours, whereas in the arctic regions it can take 10,000 years or more. This movement of water on the Earth is known as the *hydrologic*, or water cycle, and is the most important of nature's cycles, for without the transport of huge quantities of water on the land, there would be no life as we know it.

After storms bring water to the Earth's surface, what becomes of this water is of vital importance. Every year, the continents receive about 25,000 cubic miles of water. About 15 percent of the moisture in the atmosphere comes from the land, and some 15,000 cubic miles of water is evaporated from lakes, rivers, aquifers, soils, and plants. Plants lose a considerable amount of water through the process of transpiration. (Water evaporates from plant leaves to cool them and to bring nutrients in the soil up to the limbs by capillary action.) The remaining 10,000 cubic miles of water is runoff by rivers and streams, which are the most apparent part of the water cycle. Much of this surplus water is lost by floods, which in many cases, are man-made disasters.

In the United States, there are about 3 million miles of rivers and streams, and about 6 percent of the total land surface are lands adjoining these water courses that are prone to flooding. Floods are natural, recurring events and are important for the distribution of soils over the land. During a flood, a river might change its course as it meanders its way to the ocean, and in so doing, carve out a new floodplain. Floodplains provide level ground and fertile soil, but they also attract commerce, and rapid development of these areas without due consideration for the flood potential usually ends in disaster.

Since 1936, the American federal government has spent about $10 billion on flood-protection projects. Most of them are man-made reservoirs (FIG. 10-5), which even out the flow rates of rivers by having a storage capacity that can absorb increase flow during a flood. The dams also provide hydroelectrical power, which is the cleanest form of energy and which uses a renewable resource, falling water. However, reservoirs also place valuable land areas under water. Without proper soil-conservation measures, the accumulation of silt by erosion can severely limit the life expectancy of a reservoir.

To terrestrial life, soil moisture is the most important source of water since plants are a basic component of the food chain. A substantial amount of rainwater percolates into the soil where roots can draw it up into the plant. Different types of soils have different holding capacities for water, and generally the coarser, more sorted soils hold less water than fine silts and clay mixed with humus. Most of the eastern third of the United States receives sufficient rainfall to support agriculture without the need for irrigation. In contrast, much of the western portion of the nation has a rain deficit, which must be made up by irrigation.

Over 10 percent of the world's cultivated land is irrigated, requiring almost 1,000 cubic miles of water. With irrigation, crops are not dependent on the whims of nature to provide them with water, more land can be brought under cultivation, and two or more crops can be grown in a single year. The disadvantages are that overuse of groundwater for irrigation depletes aquifers and causes subsidence, which severely limits the recovery of groundwater systems. The use of river water for irrigation also has its drawbacks, particularly because of its salt content. Poorly drained fields allow salt to build up in the soil, and thousands of once-productive acres are ruined in this fashion every year.

MARINE CIRCULATION

The ocean's ability to store and move about vast quantities of heat is much greater than the atmosphere's, and this has a profound effect on the climate. The close association between the atmosphere and the ocean as well as the ocean's large influence on the weather, has only recently been recognized. The ocean's large heat capacity allows it to retain the summer's heat and slowly give it up during the winter, which moderates the Earth's temperature during the seasons. It takes about a decade to significantly change the temperature of the upper 1000 feet of the ocean, and to change the temperature of the abyssal could take 1,000 years or more.

The oceans have both surface and abyssal currents, which play a vital part in moving the Earth's heat around. The surface currents (FIG. 10-6) are driven by the wind, and as with flowing air masses, they are deflected by the Coriolis effect to the right in the Northern Hemisphere and to the left in the Southern Hemisphere. These ocean currents are similar in function to currents in the atmosphere; they transport warm water from the tropics, distribute it to higher latitudes, and return with colder water. However, they do so at a much slower pace than air currents, taking up to a decade or so to complete the journey.

Abyssal currents are driven by thermal forces whereby cold water in the polar regions descends, spreads out upon hitting the ocean floor, heads toward the equator, and dissipates in the tropics. The path taken by the deep-water currents is controlled by the distribution of landmasses and by the topography of the ocean bottom, such as ocean ridges and canyons. Abyssal currents flowing toward the equator are deflected to the west as a result of the Earth's eastward rotation, which presses the currents against the eastern edges of the continents.

(Photo by Edward B. Trovillion, courtesy of Soil Conservation Service)

FIG. 10-5. Lake formed by flood control dam.

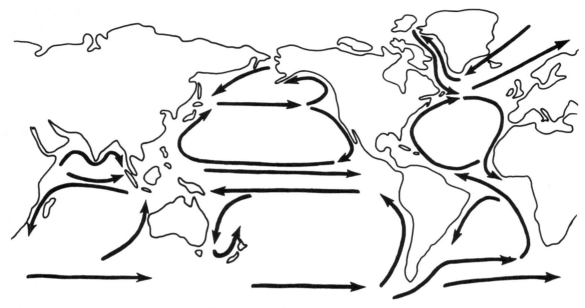

FIG. 10-6. Major ocean currents.

The western boundary undercurrent off eastern North America transports some 20,000 cubic miles of water yearly–twice the amount of water that is carried by all the world's rivers and streams.

The sinking of water in the poles is matched by upwelling currents in the tropics, creating an efficient heat–transport system that takes upwards of a thousand years to complete a single cycle. Upwelling currents are most important in transporting ocean-bottom nutrients to the surface to support sea life. Although these zones might cover only 1 percent of the ocean's surface area, they sustain about 40 percent of all marine life.

There are storms in the ocean just like there are storms in the atmosphere. Because water is a more dense fluid medium than air, the currents tend to be much slower but are still very influential forces. Some parts of the ocean currents are pinched off, forming eddies, or rings of swirling water, like underwater hurricanes, as much as 100 miles or more across and as deep as 3 miles below the surface. Most of these rings are less than 50 miles across and play an important role in mixing the ocean waters. Marine life are often trapped in the rings and transported to hostile environments, where they live only as long as the rings do, which can take from several months to over a year. Abyssal storms, which last anywhere from several days to a few weeks, transport huge loads of sediments and play an important role in modifying the seafloor, giving the ocean bottom a much more complex marine geology than it was once thought. The currents only move about 1 mile per hour but can scour the ocean floor just as effectively as an offshore gale with heavy winds. Most of the detritus that is redistributed consists of shells and skeletons of dead marine organisms that rain down on the ocean floor, along with sediment washed off the continents.

Some ocean currents have a dramatic effect on the weather, and changes in these systems can send abnormal weather patterns all around the world. One such current is an annual southerly flow of warm water off the west coast of South American that begins around Christmas and leaves by Easter. It is named *El Niño*, Spanish for the Christ child. The current temporarily disrupts the upwelling of nutrient-rich cold water, which adversely affects the local fishing industry.

Table 10-1. History of the Deep Circulation in the Ocean.

AGE	EVENT
> 50 million years ago	The ocean could flow freely around the world at the equator. Rather uniform climate and warm ocean even near the poles. Deep water in the ocean is much warmer than it is today. Only alpine glaciers in Antarctica.
35–40 million years ago	The equatorial seaway begins to close. There is a sharp cooling of the surface and of the deep water in the south. The Antarctic glaciers reach the sea with glacial debris in the sea. The seaway between Australia and Antarctica opens. Cooler bottom water flows north and flushes the ocean. The snow limit drops sharply.
25–35 million years ago	A stable situation exists with possible partial circulation around Antarctica. The equatorial circulation is interrupted between the Mediterranean Sea and the Far East.
25 million years ago	The Drake Passage between South America and Antarctica begins to open.
15 million years ago	The Drake Passage is open; the circum-Antarctic current is formed. Major sea ice forms around Antarctica, which is glaciated, making it the first major glaciation of the Modern Ice Age. The Antarctic bottom water forms. The snow limit rises.
3–5 million years ago	Arctic glaciation begins.
2 million years ago	An ice age overwhelms the Northern Hemisphere.

About every 5 to 8 years, anomalous atmospheric pressure changes in the South Pacific, called the *Southern Oscillation*, causes the westward-flowing trade winds to collapse, and warm water piled up in the western Pacific by the winds flows back to the east, creating a great sloshing of water in the South Pacific Basin. This sloshing thickens the layer of warm water in the eastern Pacific (FIG. 10-7), which suppresses the *thermocline*, the boundary between cold and warm layers of seawater, and prevents the upwelling of cold water from below, thus creating optimum conditions for an El Niño event.

The El Niño warming event of 1972-73 devastated the Peruvian anchovy industry, which fell from an annual catch of more than 13 million tons in 1970 to less than 0.5 million tons in 1983. The El Niño of 1982-83–was remarkable for the amount

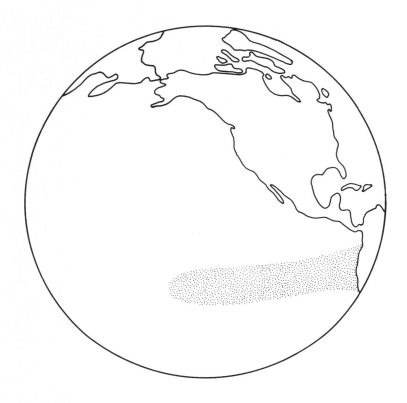

FIG. **10-7.** Extent of El Niño warming in the Pacific.

of property damage and human suffering it caused worldwide, and will go down in the annals as one of the most tragic climatic events in human history.

THE BIG MIX-UP

The mixing of temperature and gases between the atmosphere and the ocean is mostly conducted by wave action. Waves are produced by wind blowing across the surface of the water and can range from small ripples to giants 100 or more feet high. Storms at sea are responsible for the constant beating of waves against a beach, and offshore hurricanes produce a storm surge that can be extremely damaging.

The wind-stirred layer of the ocean, which involves the upper 300 feet, is the most uniform environment on Earth and is in equilibrium with the atmosphere at all times. Within this layer, the *phototropic zone* , exists most of the marine biomass, which must be near the surface in order for the Sun

to penetrate for photosynthesis. The surface action of the ocean plays an important role in the exchange of carbon dioxide and oxygen, of which 80 percent of the Earth's total supply is generated by marine photosynthesizers.

Another transfer mechanism from the sea to the air involves the transport of marine substances by the wind. These substances are ejected into the atmosphere by bursting air bubbles and ocean spray from waves. The spray evaporates into small particles of sea salt that are wafted aloft by air currents. Upwards of 10 billion tons of salt enter the atmosphere in this manner every year. The salt provides seed crystals for the condensation of rain, as well as nutrients for coastal areas and high latitudes above the vegetation line.

Breaking waves dissipate energy along the coast and are responsible for generating along-shore currents, which transport sand along the beach. They also cause coastal erosion, a serious problem in many coastal areas where the shoreline is receding at alarming rates. Most of the high waves and beach erosion occur during coastal storms, mostly thunderstorms and squalls. Hurricanes with winds upwards of 100 miles per hour and more produce the most dramatic storm surges, which are responsible for destroying beach-front properties.

Beach erosion is very difficult to predict and almost impossible to stop. It is influenced by the strength of beach dunes or sea cliffs, the intensity and frequency of coastal storms, and the exposure of the coast. Most attempts to halt beach erosion fail because waves constantly batter and erode man-made defenses. Jetties and seawalls erected to prevent further damage actually exacerbate erosion by cutting off the supply of sand or by reflecting the waves and the sand back out to sea rather than allowing the beach to absorb wave energy. The results are vanishing barrier islands from Virginia to east Texas and homes in California falling into the sea.

Tidal waves, more correctly termed *seismic sea waves* or *tsunamis*, really have nothing to do with the tides, and are instead produced by undersea earthquakes, volcanoes, or landslides. The undercutting of a sea cliff through wave action and excessive rainfall along the coast can cause huge blocks to slide into the sea and produce a tsunami.

The seismic sea waves from a powerful earthquake travel at great speed, practically unnoticed on the surface of the ocean. When the waves drag on the bottom of a steeply rising shore, the resistance causes them to abruptly slow down. This sudden breaking action forces the water to pile up on itself, and the wave immediately grows into a giant. Its destructive force is immense: large buildings are crushed with ease, and ships are carried well inland as though they were toys.

Generally, coastal residents have no warning of the impending disaster, except for a rapid withdrawal of water from the beach before the wave strikes. This lack of warning has prompted the installation of a tsunami warning system, which responds to earthquakes of magnitude 7.5 or better on the Richter scale.

By far, the most tsunami-prone region of the world is the rim of the Pacific Ocean. Because of crustal collisions, this region has the most earthquakes as well as the most volcanoes.

UNDERWATER VOLCANOES

Volcanic eruptions on the Earth's surface, the ocean floor, and the midocean ridges act as vents to exhaust the Earth's interior heat and to replenish chemical substances the Earth needs to stay alive. More than 80 percent of the Earth's surface above and below sea level is of volcanic origin, and most of the volcanic activity that continually remakes the surface of the Earth takes place on the bottom of the ocean. At midocean ridges, volcanic emissions continually build new ocean crust, which carries whole continents with it as it spreads apart.

Most of the world's islands began as undersea volcanoes. Successive eruptions piled up volcanic rocks until the peak poked through the water's surface. Oceanic volcanoes are among the most explosive in the world, and whole islands disappear when the volcano blows its top. Many volcanic islands form island arcs with graceful curves that are indicative of crustal movements. Volcanic ash makes a rich soil, and before long, seeds carried by the wind, sea, and even birds will turn a newly formed island into a lush tropical paradise, replete with all the necessities of life.

11

The Supercontinent

CONTINENTAL drift had a wide-ranging effect on the distribution and isolation of species. The changes in continental configuration greatly affected global temperatures, ocean currents, productivity, and many other factors of fundamental importance to life on Earth. The positioning of the continents helped determine climatic conditions. When most of the land huddled near the equatorial regions, the climate was warm, but when lands wandered into the polar regions, the world became covered with ice. Taking land from the tropics and replacing it with ocean also had a net cooling effect because land in the tropics absorbs more of the Sun's heat than oceans do. When Antarctica drifted over the South Pole and the Arctic Ocean became enclosed by land, warm ocean currents were blocked from the polar regions. This blockage caused ice to form at both poles—a rare event in the Earth's history. Since then, ice ages have come and gone almost like clockwork.

The changing shapes of ocean basins as a result of the movement of continents affects the flow of ocean currents, the width of continental margins, and consequently, the abundance of marine habitat. During times of high active continental movements, there is also greater volcanic activity, especially at midocean ridges. The amount of volcanism could affect the composition of the atmosphere and ocean, the rate of mountain building, the climate, and inevitably, life itself.

PANGAEA

Early sixteenth century map makers must have wondered why the Atlantic coastlines of South America and Africa seemed to fit together like pieces of a global jigsaw puzzle. There were similar rock types and fossils on continents widely separated by ocean. There were also many similarities between the living plants and animals of the New World and the Old.

Several theories were put forth to explain this paradox. Some seventeenth and eighteenth century geologists argued that the Great Flood gouged out a giant valley that flooded with the advancing sea, dividing the continents with matching shorelines. It was once thought that the Moon was plucked out of the Earth, leaving a great rent in the Earth's crust which created the Pacific Basin. As this filled with magma from the Earth's interior, the continents floated toward the basin like rafts on a river of molten rock.

Another idea was that the ocean basins developed from sinking land. It was even thought that the Earth was shrinking and that blocks of crust collapsed to fill the voids. It was reasoned that the continents were connected by land bridges, which later sunk beneath the ocean, but meanwhile, they gave animals a convenient pathway to travel from one part of the world to another. It seemed highly unlikely that such a variety of species could have evolved along parallel lines without some means of crossing from one continent to the next.

The idea of the existence of a single large continent in the geologic past did not come from one individual, but was rather a conclusion developed independently by several investigators after study of much data that provided overwhelming evidence for such a possibility. Fossil bones of a 2-foot-long Triassic reptile called lystrosaurus were found in South Africa, Antarctica, India, and China (FIG. 11-1). Its existence in such diverse places that are now widely separated by ocean was hailed as proof for the existence of Gondwanaland, since it was very unlikely that the animal could swim across such a great distance.

The Late Paleozoic fern glossopteris (mentioned in Chapter 9) was found everywhere in the Southern Hemisphere, but was suspiciously missing in the Northern Hemisphere, suggesting that at one time, two large landmasses were once separated by a sea. Scientists named this sea the Tethys. Similar geological provinces with the same rock types in the same succession were found on continents presently separated by thousands of miles of water (FIG. 11-2).

The *magnetic inclination*, or vertical flux of the Earth's magnetic field, imprinted on basalts as they cooled indicated that the continents moved relative to the magnetic poles. All the southern continents show evidence of contemporaneous glaciation during the Upper Paleozoic, and some large boulders

carried by the glaciers only matched with rocks on the opposite continent. The continents also have coal deposits with similar fossilized plant material, and coal has been found in the polar regions which indicate a once-tropical climate.

In 1915, the German meteorologist and arctic explorer Alfred Wegener proposed the theory of continental drift based on crustal movements. Accordingly, about 200 million years ago, all the landmass was consolidated into a supercontinent Wegener called *Pangaea*, meaning "all lands," and all the oceans were combined into a single body of water which he called *Panthalassa*, meaning "universal sea." Wegener supported his theory with an impressive collection of evidence, including the geometric fit of continental margins, matching mountain ranges on opposite continents (FIG. 11-3), corresponding rock successions, similar ancient climatic conditions, and fossils of identical species. Crustal movements also provided a better method for explaining mountain building (FIG. 11-4), but Wegener incorrectly thought that mountains were formed when the leading edge of continents crumpled as the landmass plowed its way through the oceanic crust, like an ice breaker cuts its way through pack ice.

Wegener's inability to provide an adequate mechanism for moving the continents around was enough for most scientists of his day to reject his theories out of hand. Unfortunately, Wegener did not live to see his continental drift theory gain acceptance by the scientific community and become incorporated into a broader field called *plate tectonics*.

THE BIG CRACK

It was only after geological and geophysical data from the oceans became so overwhelmingly compelling that scientists were forced to abandon the archaic thinking of the past century. By the late 1960's, most Northern Hemispheric geologists joined their southern colleagues, who had been convinced for some time of the reality of continental drift.

The generally accepted picture is that the present continents were sutured together into the supercontinent of Pangaea during the Permian and

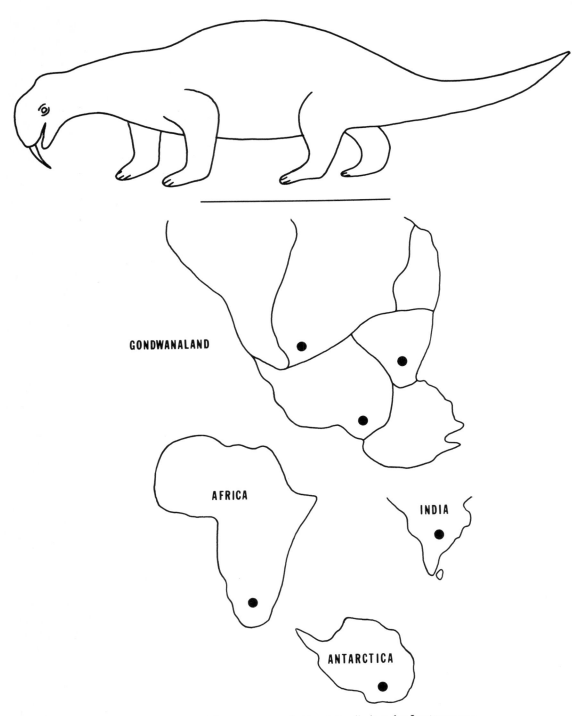

FIG. 11-1. (top) Lystrosaurus; (bottom) fossil sites for Lystrosaurus.

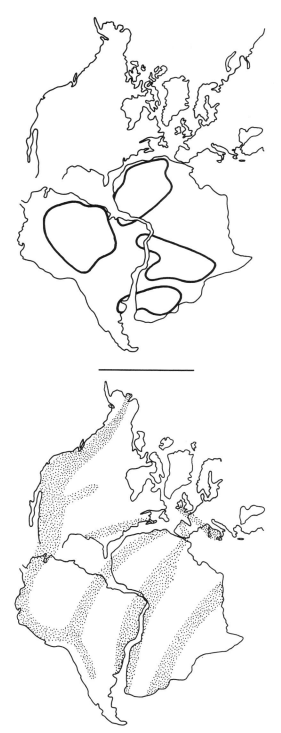

FIG. 11-2. (top) Geological provinces and the fit of the continents. (bottom) Worldwide tin belts and the fit of the continents.

Lower Triassic periods. In the Upper Triassic and Jurassic, Pangaea began to break apart along a fracture zone now represented by the Mid-Atlantic Ridge. The breakup finally was completed sometime during the Early Cretaceous, and since that time, all continents bordering on the Atlantic have been drifting away from each other.

The split-off of Antarctica, Australia, and India was a little more complex, and they each took different paths. About 50 million years ago, India rammed into Eurasia, and the underthrusting of the Indian plate raised the 300,000-square-mile Tibetan plateau. Also, Alaska and Australia once might have had a lot in common. It seems that some 500 million years ago, a large portion of the Alaskan panhandle was part of eastern Australia. It broke off and drifted across the Pacific about 375 million years ago and bumped into North America about 100 million years ago.

Probably the only thing that kept most skeptics from accepting continental drift was the direct detection of continental movements. Presently, the Atlantic basin is widening and spreading North America and Europe apart at a rate of about 1 inch per year. This rate was probably much greater during the early breakup of Pangaea. Trying to measure 1 inch over the distance of several thousand miles strains the limits of most measuring devices; however, there are some distance-measuring techniques that are remarkably precise.

If the Earth had no atmosphere to interfere with radio signals from quasars in deep space or satellite measuring systems, drift probably would be more apparent. As it is, the amount of error induced by the atmosphere can be several times greater than the drift itself. As a result, measurements must be taken over a period of several years in order to establish a baseline. A better understanding of the atmospheric interference, as well as other external effects such as irregularities in the Earth's rotation caused by the gravitational pull of the Sun and Moon, and a better accounting for systematic errors might increase accuracy significantly. Observed average plate motions by satellite laser ranging between the North American, Pacific, and Australian plates appear to be in general agreement with the geological

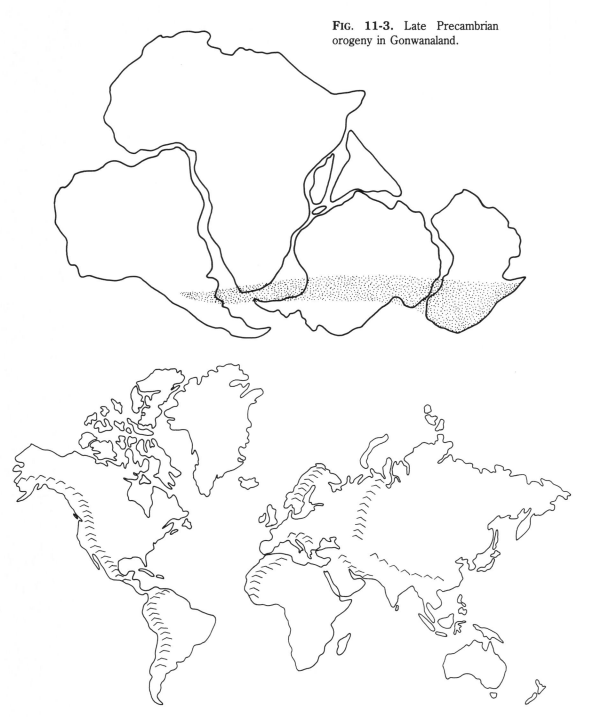

FIG. 11-3. Late Precambrian orogeny in Gonwanaland.

FIG. 11-4. Major continental mountain ranges.

rates, which are averaged over at least 100,000 years.

The geological rates were determined by paleomagnetic studies of basalts laid down on the ocean floor on either side of a spreading ridge. As lava, which is normally iron-rich, extrudes from a rift system and cools, the magnetic fields of its iron molecules align themselves to the Earth's magnetic field like tiny compasses. The geomagnetic field has reversed itself hundreds of times over the past 200 million years. If the Atlantic seafloor were spreading apart, successive layers of basalt should be found with alternating magnetic fields going outward from the Mid-Atlantic Ridge. To test this theory, *magnetometers*, which are sensitive magnetic recording instruments, were towed behind ships traversing the ridge system. They produced a remarkable picture of magnetic stripes of varying width and magnitude with identical successions of reversing magnetic fields on both sides of the ridge.

The magnetic reversals provided a convenient means of dating the ocean floor because no sequence of reversals followed the same pattern and therefore, were not periodic. Finding the rate of spreading was a simple matter of dating the magnetic stripes and measuring the distance from their points of origin.

Now that it seems fairly certain that continents are moving relative to each other, the question remains how the continents broke apart in the first place. The *lithosphere*, which includes the crust and its underlying plate, is generally between 50 and 100 miles thick on the continents, whereas on the ocean floor, it ranges from about 5 miles thick near spreading centers to no more than 60 miles thick at plate margins. Cracking open the continental lithosphere therefore seems to be a mighty task. Somehow, when continental lithosphere rifts into two separate plates, thick lithosphere must give way to thin lithosphere.

The best evidence for the rifting of continents can be found in the East African Rift System (FIG. 11-5), which has not yet fully ruptured. When this occurs, the continental rift will be replaced by an oceanic rift, such as what is presently taking place in the Red Sea.

FIG. 11-5. The East African rift system.

This transition is accompanied by *block faulting*, whereby blocks of continental crust drop down along extensional faults, resulting in a deep rift valley and a thinning of the crust. Upwelling of molten rock from the mantle further weakens the crust and convection currents pull the crust apart. Therefore, the engine that drives the birth and evolution of rifts, and consequently the breakup of continents and the formation of ocean basins, ultimately comes from the mantle, which is constantly churning over hot rocks in a vast fiery cauldron.

THE WIDENING GAP

When Pangaea first started to break apart, some 165 million years ago, a rift began in the present-day Caribbean and sliced its way northward through the continental crust connecting North America with northwest Africa and Eurasia, and began to open up the Atlantic Ocean (FIG. 11-6). The process took several million years along a zone that was several hundred miles wide. By about 125 million years ago, the infant North Atlantic Ocean had an active mid ocean ridge system that began to make new oceanic crust. At the same time, a rift began to separate South America from Africa as they moved together away from North America and Eurasia. About 80 million years ago, the North Atlantic became a full-

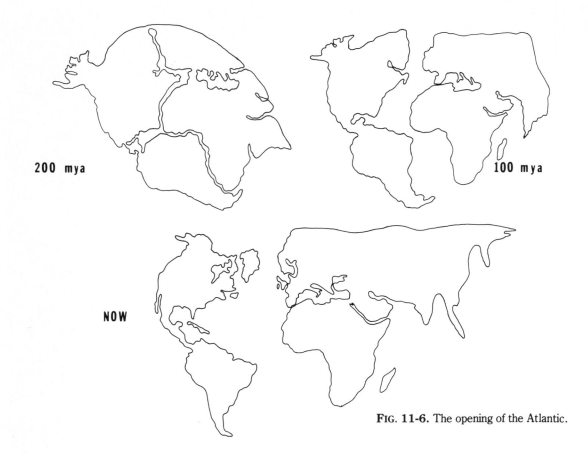

200 mya

100 mya

NOW

FIG. 11-6. The opening of the Atlantic.

fledged ocean, and North America and Greenland began to drift apart.

It was at this time that the South Atlantic Ocean began to develop, opening up from south to north like a zipper, roughly 50 million years later than its northern counterpart. This rupture produced a huge, continuous submerged mountain range surpassing in scale the Alps and Himalayas combined. The Mid-Atlantic Ridge, which has a mean depth of about 1.5 miles, winds its way down the middle of the ocean from Iceland to Bouvet Island roughly 1000 miles off the coast of Antarctica. The Mid-Atlantic Ridge closely matches the contour of opposing coastlines.

The floor of the Atlantic is a huge conveyor belt, transporting lithosphere from its point of origin at the Mid-Atlantic Ridge to its final destination down the deep-sea trenches. The ocean floor at the crest

of the ridge consists mostly of basalt. As the distance from the crest increases, the bare rock is covered with an increasing thickness of sediments, mostly red clay from detritus washed off the continents and wind-blown dust.

Near the ridge crest, these sediments are predominantly composed of calcareous ooze built up by a rain of decomposed skeletons of microorganisms from above. Farther from the crest, the slope falls below the *calcium-carbonate compensation zone*, the depth at which calcareous sediments dissolve in seawater. Therefore, in deep water far from the crest of the midocean ridge, only red clay should exist. Yet cores taken from the abyssal plains near continental shelves, where the oceanic lithosphere is the oldest, have shown thin layers of calcium carbonate below thick sequences of red clay. Apparently, the red clay protected the calcium carbonate that originated at the midocean ridge from being dissolved in seawater. This evidence implies that the midocean ridge was the source of the calcium carbonate discovered near continental margins and is strong evidence for seafloor spreading.

The theory of seafloor spreading was proposed by the American geologist Harry Hess in 1962 after he observed sonographs of a system of ridges that encircled the Earth like the stitching on a baseball (FIG. 11-7). The undersea mountains were crisscrossed at right angles by transform faults (FIG. 11-8), which resulted from lateral strain as though the ocean floor were moving sideways in relation to the midocean ridges. In addition, the midocean ridges were the sites of frequent earthquakes and volcanic activity, which formed new islands. This activity seemed to be more intense in the Atlantic where the ridge was steeper and more jagged, giving every indication that the ridge was adding new material to the ocean floor. Strange undersea vol-

canoes in the Pacific, called *guyots* (pronounced ghee-ohs), were at one time above sea level, but constant wave action eroded them below the surface as though the tops of the cones were sawed off.

The remarkable thing about these volcanoes is that the farther they were from volcanically active regions of the ocean, the older and squatter they became, which suggests that the guyots wandered across the ocean floor far from their places of origin. In this respect, the islands were probably produced assembly-line fashion, with each one moving away in succession from a magma chamber, called a *hot spot*, lying just beneath the ocean crust. Presently, the main island of Hawaii has the only active volcanoes in the Hawaiian chain with Kilauea erupting periodically since 1983.

The process of seafloor spreading begins when hot rocks rise up by convection currents in the upper mantle. Upon reaching the underside of the lithosphere, they spread out laterally, cool, and descend back into the Earth's interior, completing a

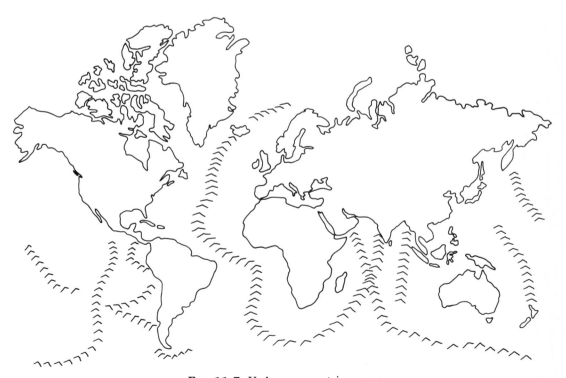

FIG. 11-7. Undersea mountain ranges.

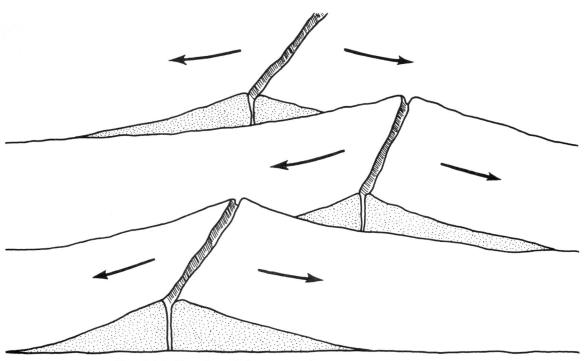

FIG. 11-8. Transform faults at spreading centers on the ocean floor.

single convection loop in perhaps 100 million years or so. The constant pressure against the bottom of the lithosphere causes fractures to form and forces it to open up. As the convection currents flow out on either side of the fracture, they carry the now separated parts of the lithosphere with them, and the opening widens. The pressure is thus reduced and mantle rocks melt and rise through the fracture zone, where the molten rock finds easy passage through the 60 miles or so of lithosphere until it reaches the crust. There, it forms magma chambers that further press against the crust and widen it, simultaneously providing magma that pours out from the trough between ridge crests, and adds layer upon layer of basalt to both sides of the spreading ridge. The pressure of the upwelling magma forces both sides of the ridge farther apart and pushes the ocean floor and the lithospheric plate on which it rides away from the midocean ridge. In this manner, one or more continents are also carried passively along on about a half dozen major lithospheric plates created at the spreading centers and destroyed in the deep-sea trenches.

DEEP-SEA TRENCHES

In the 1870s, the British oceanographic ship H.M.S. *Challenger* was taking samples off the Marianas Islands in the western Pacific when it encountered a huge trench at a depth of about 5 miles. The deep-sea trenches off continental margins and island chains (FIG. 11-9) were thought to be great bulges in the ocean crust, called *geosynclines*, that were caused by the weight of tremendous amounts of sediments washed off the continents. The trenches were later found to be regions of intense earthquake and volcanic activity. Volcanic island arcs fringed the trenches, and each had similar curves and similar origins. Since the Earth was not expanding, the new ocean crust created at the midocean ridges had to disappear somewhere, and the deep-sea trenches appeared to be the most likely places.

In 1954, the American seismologist Hugo Benioff discovered a descending plate by studying deep-seated earthquakes that acted like beacons, outlining the boundaries of the plate. The lithosphere entered the Earth's interior at an angle of about 45 degrees just as it should according to theory. Upon reaching the boundary between the lower and upper mantle about 400 miles beneath the surface, the plate was prevented from descending any farther and in time, became incorporated into the general circulation of the mantle.

It was generally thought that the push the plates received from seafloor spreading would be sufficient to force them to dive into the mantle. Apparently, however, the drag forces at the base of the plates can greatly resist plate motion, and some other force is also needed. Therefore, the force of gravity is looked upon as the driving mechanism and pull is favored over push in order to overcome the plate drag.

The farther plates expand from the spreading centers, the thicker and colder they become. By the time the plates reach the deep-sea trenches, they have cooled so much since their formation that they are dense enough to sink into the mantle at the deep-sea trenches, also called *subduction zones*. The pull on the ocean plate sinking into the mantle by convection currents might provide additional force to overcome the resistance. The magnitude of the pull of the sinking plate would depend on the length of the subduction zone, the rate of subduction, and the amount of trench suction. With these forces in place, the plates could practically drive themselves without the need for driving forces from seafloor spreading.

If you follow the course of a descending plate, you would see that it is first subjected to internal stresses, causing it to fault as it is forced to bend and dive under another plate. As the plate begins its passage through the mantle, its temperature rises slowly at first, mainly because of internal radiogenic heat sources. Because the heat cannot escape from the plate as it did when it was near the surface, the

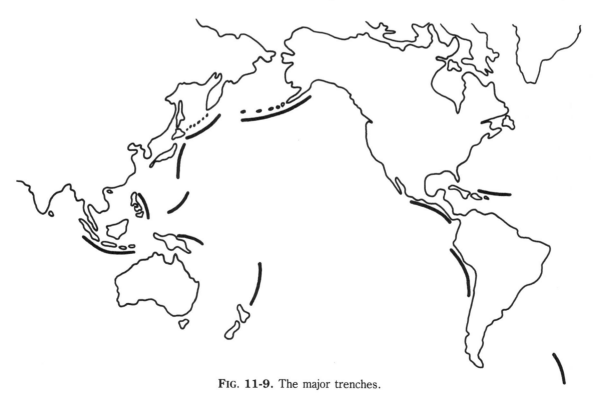

FIG. 11-9. The major trenches.

temperature steadily builds up. As the plate dives deeper into the mantle, heat is eventually conducted from the outside, where the plate is in contact with the mantle. The plate is also subjected to increasing pressure as it passes through the mantle. By the time the plate reaches a depth of about 375 miles, the extreme temperatures and pressures transform it into a highly dense material with a compact crystal structure. As the plate continues downward, the tightly packed crystals begin to partially melt, and the rocks become plastic and are able to flow. When the subducted segment of the plate reaches the boundary of the upper and lower mantle, it finds its downward thrust blocked, and it must bend along the boundary and move in the direction of the convective flow.

The entire trip from top to bottom could take some 10 million years. In another 50 million years the plate would become totally indistinguishable from the mantle; in other words, the plate would be completely consumed by the mantle.

DRIFT AND THE HISTORY OF LIFE

It is becoming more apparent that plate tectonics and continental drift have been operating since the early history of the Earth and have played a prominent role in the history of life. Changes in the relative configuration of the continents and the oceans in all likelihood had a far-reaching influence on the environment, climate patterns, and the composition and distribution of life in the biosphere. The continual changes in world ecology had profound effects on the course of evolution and accordingly, on the diversity of living organisms.

The chief process by which evolution proceeds is *natural selection*, which is essentially an ecological process based on the relationship between organisms and their environment. Certain inherited traits allow species to become particularly well suited to survive and reproduce in their prevailing environment. During environmental stress, species that acquire favorable traits through mutations adapt more easily than other species and are more likely to survive and pass on their survival characteristics to their offspring. Because there are a large number of different environments, the result is a wide variety of spe-

cies. Therefore, evolutionary trends varied throughout geologic time in response to major environmental changes as natural selection acted to adapt organisms to the new conditions forced on them by a number of environmental factors brought about by, among other things, continental drift.

When Pangaea was welded into one large continent around 230 million years ago, the Earth had a great diversity of plant and animal life on the land and in the sea. The large landmass near the tropics allowed more of the Sun's heat to be absorbed by the Earth, which contributed to higher global temperatures. Oceans in the high latitudes were less reflective than land and absorbed more heat, which further moderated the climate. Also, because there was no land in the polar regions to interfere with the movement of warm ocean currents, both poles remained ice free year round and the temperature did not vary greatly between the high latitudes and the tropics.

During the breakup of Pangaea, the climate of Earth, particularly in the Cretaceous period, was extremely warm; global temperatures averaged 10 to 25 degrees Fahrenheit warmer than they are today. When the continents drifted toward the poles by the end of the Cretaceous, they disrupted the transport of poleward oceanic heat and substituted reflective, easily chilled land for absorptive, heat-retaining water. As the cooling progressed, the land accumulated snow and ice, creating an even greater reflective surface and further lowering global temperatures.

The vast majority of marine species live on continental shelves or shallow-water portions of islands and subsurface rises generally less than 600 feet deep (FIG. 11-10). The richest shallow-water faunas are in the tropics, which are crowded with large numbers of highly specialized species. As the latitude gets higher, diversity gradually falls off until in the polar regions, there are less than one-tenth as many species as in the tropics.

The diversity depends mostly on the food supply. As the seasons become more pronounced in the higher latitudes, there are greater fluctuations in food production. Diversity is also affected by seasonal changes, such as variations in surface and upwelling currents that affect the nutrient supply,

FIG. 11-10. The distribution of shelf faunas.

which in turn causes large fluctuations in productivity. Therefore, the greatest diversity among species is off the shores of small islands or small continents in large oceans where fluctuations in the nutrient supply are least affected by seasonal effects of landmasses. Diversity is also highly dependent on the shape of the continents, the width of shallow continental margins, the extent of inland seas, and the presence of coastal mountains, all of which are affected by continental motions.

When all the continents were assembled into the supercontinent of Pangaea at the beginning of the Triassic period, a continuous shallow-water margin ran all the way around it with no major physical barriers to the dispersal of marine life. Also, during that time, the seas were largely confined to the ocean basins and did not extend significantly over continental shelves. Consequently, habitat area for shallow-water marine organisms was very limited, thus accounting for low species diversity. As a result, marine biotas in the Triassic were more widespread but

were made up of comparatively fewer species.

Similar circumstances might have occurred in the Late Precambrian period, when another supercontinent appears to have been in existence. During the Cambrian, it broke apart into four separate continents, which probably had a major effect on the explosion of species during that time. When Pangaea broke up and the resulting continents migrated to their present positions, diversity again increased to unprecedented heights, providing a rich variety of species.

Since it is apparent that continental drift had and continues to have a major impact on the evolution of life on Earth, the question remains how future configurations of the continents (FIG. 11-11) will affect life. The Atlantic basin will continue to expand at the expense of the Pacific basin as North and South America head westward. The Panama isthmus connecting the Americas will sink out of sight as the two continents continue to pull apart, allowing free passage for currents and sea life from the Atlantic

Table 11-1. The Drifting of the Continents.

AGE (IN MILLIONS OF YEARS)		GONDWANALAND	LAURASIA
Quaternary	3		Opening of Gulf of California
Pliocene	11	Begin spreading near Galapagos Islands Opening of the Gulf of Aden	Change spreading directions in eastern Pacific Birth of Iceland
Miocene	25	Opening of Red Sea	
Oligocene	40	Collision of India with Eurasia	Begin spreading in Arctic Basin
Eocene	60	Separation of Australia from Antarctica	Separation of Greenland from Norway
Paleocene	65	Separation of New Zealand from Antarctica Separation of Africa from Madagascar and South America	Opening of the Labrador Sea Opening of the Bay of Biscay Major rifting of North America from Eurasia
Cretaceous	135	Separation of Africa from India, Australia, New Zealand, and Antarctica	
Jurassic	180		Begin separation of North America from Africa
Triassic	230		
Permian	280		

FIG. 11-11. Drift of the continents 50 million years from now.

to flow into the Pacific. Africa and Eurasia will continue to press against each other, and the Mediterranean Sea, caught in the middle, will be squeezed dry while new mountains rise up on either side. The Arabian subcontinent will break free from East Africa, as Madagascar had done earlier, and collide with India. Australia will continue to drift northward and crash into Southeast Asia. Eventually, all the continents will participate in another great continental collision, creating a new supercontinent called *Neopangaea*. As before, species diversity will suffer until such a time when the supercontinent breaks up again. Then there will be another explosion of species, which will bear little resemblance to life on Earth today.

12

✤

Life in Strange Places

LIVING organisms occupy just about every conceivable environment on Earth, from subfreezing to boiling and everything in between. Life on the surface of the Earth is so apparent that people often forget most of the world's biomass is hidden out of sight. The oceans contain some 80 percent of the biomass and most of the species of plants and animals. In the depths of the abyssal region, many animals exist in the cold and the dark and have adapted to such high pressures that they do not live for very long if taken to the surface.

On the ocean floor near the East Pacific Rise, strange deep-sea animals live in communities clustered around hydrothermal vents that provide both warmth and nutrients. It is the only environment on Earth that is totally independent of the Sun for its energy. This energy comes instead from the interior of the Earth. Hydrothermal vents on the Earth's surface and on the deep-sea floor provide mineral-rich hot water in which bacteria flourish at temperatures hot enough to boil an egg. Arctic fishes are able to live in subfreezing water beneath the ice because of an antifreezelike substance in their blood

that keeps their bodily fluids from freezing. Underground, bacteria break down organic materials, which help enrich the soil and aid in nitrogen fixation for plants. Worms and insects till the soil, providing much needed organic matter essential to growing plants, which are ultimately the basis for all life on the land.

LIFE IN THE DESERT

Deserts are not only the hottest and driest regions, but many deserts are also the most barren environments on Earth. Only the hardiest of plant and animal species live there, including plants whose seeds can survive a 50-year drought and rodents that live their entire lives without taking a single drink of water, surviving off the water generated solely by their metabolism. Technically, a *desert* is defined as an area that receives less than ten inches of precipitation annually. Much of the world's desert wasteland receives only minor quantities of rain during certain times of the year, while other areas have gone totally without rain for several years.

Table 12-1. Major Deserts of the World.

DESERT	LOCATION	TYPE	AREA (SQUARE MILES × 1000)
Sahara	North Africa	Tropical	3500
Australian	Western/interior	Tropical	1300
Arabian	Arabian Peninsula	Tropical	1000
Turkestan	S. Central U.S.S.R.	Continental	750
North America	S.W. U.S./N. Mexico	Continental	500
Patagonian	Argentina	Continental	260
Thar	India/Pakistan	Tropical	230
Kalahari	S.W. Africa	Littoral	220
Gobi	Mongolia/China	Continental	200
Takla Makan	Sinkiang, China	Continental	200
Iranian	Iran/Afghanistan	Tropical	150
Atacama	Peru/Chile	Littoral	140

About 30 percent of the Earth's land surface is classified as desert (FIG. 12-1). Because of natural and human activities, more and more land is becoming desertified—some 15,000 square miles, or about the size of the Mojave Desert of California, each year. Only about 10 percent of the desert lands are composed of sandy dunes, which march across the desert floor in response to the winds (FIG. 12-2).

Perhaps the most impoverished desert on the face of the Earth is found surprisingly in Antarctica (FIG. 12-3). Dry valleys running between McMurdo Sound and the Transantarctic Mountains receive less than four inches of snowfall per year, most of which is blown away by strong winds. Only meager signs of life are found: blue-green algae on the bottom of small, glacier-fed lakes; bacteria in the soil; and a wingless fly—the largest land animal on the continent.

Most of the world's great deserts exist in the subtropics in a broad band running between 15 and 40 degrees latitude north and south of the equator. In the Northern Hemisphere, a series of deserts stretch from the west coast of North Africa through the Arabian peninsula and Iran and on into India and China. In the Southern Hemisphere, a band of deserts runs across South Africa, central Australia, and west-central South America.

After the precipitation has been wrung out of the rising air over the tropics, little moisture is carried by upper air currents to the subtropics. The dry air cools and sinks, producing zones of semipermanent high pressure, called *blocking highs*, because they block advancing weather systems from entering the region, producing clear skies and calm winds most of the time. Tall mountains also block weather systems. Generally on the lee side of a mountain is an area called a *rain shadow zone* where only small amounts of precipitation fall because the mountains force clouds to rise higher, which causes all the precipitation to occur on the windward side.

Since deserts are generally light-colored, they have a high albedo, so they reflect large amounts of solar energy back into space. Desert sands also absorb a great deal of heat during the day; the surface can scorch at temperatures exceeding 150 degrees Fahrenheit. However, because the skies are mostly clear during the night, any thermal energy trapped in the sand quickly escapes because of its low heat capacity, making the desert cold at night with temperatures dropping near freezing. As a result, desert regions have the highest daily temperature extremes of any environment on Earth.

The adaptations needed by plants and animals to survive in the desert generally rely on the con-

FIG. **12-1.** Major world deserts.

FIG. **12-2.** Sand dunes at Death Valley, California.

servation of water, a most precious and scarce commodity, and suspended animation during the driest part of the season. Plants like the giant saguaro cactus (FIG. 12-4) store large amounts of water in their trunks, and some plants are able to extract moisture directly from the air. During the hottest part of the day, many animals retreat to underground burrows, where the temperature can be as much as 90 degrees Fahrenheit cooler. Even the space a few feet above the ground can drop as much as 60 degrees Fahrenheit, and animals perched on small bushes find the air relatively cool.

Fish and amphibians, which flourish during the rainy season, have only a short time to lay their eggs before the ponds dry up. The animals then burrow in the bottom mud and lie dormant until the rains return. During the next rainy season, the animals come back to life, the eggs hatch, and the cycle is repeated. One extreme example is the tiny fairy shrimp of the Namib Desert of southern Africa, whose eggs might lie dormant for 20 or more years. When a rare rain shower fills the dry, shallow basin where the eggs are laid, it subsequently becomes teeming with life, and the shrimp have only a short period in which to lay their eggs before the water completely evaporates, leaving the pool cracked and parched once again.

The lowest place on Earth is the Dead Sea, 1300 feet below sea level. It is located in the Syrian Desert on the border between Israel and Jordan. For thousands of years, freshwater carrying salts leached from the rocks has flowed south through the Jordan Rift Valley and terminated into the Dead Sea. Because there is no outlet, the inflowing water

(Photo by W.B. Hamilton, courtesy of USGS)

FIG. 12-3. Volcano rising above the glacial ice in Antarctica.

evaporates in the desert air, leaving salts to accumulate in the lake. As a result, the Dead Sea is not only one of the world's deepest lakes, about 1000 feet deep, but it is also the world's saltiest, with an average salinity eight times greater than the ocean. This hypersaline water is not particularly hospitable to life, hence the lake's name.

Since the 1960s, the diversion of freshwater from rivers feeding the Dead Sea for irrigation of agricultural lands has lowered the level of the lake, making its surface water even saltier and much denser and producing a highly unstable condition with a heavy, salty layer of water over a layer of lighter water on the bottom of the lake. As a result, the lake has completely turned over, with the surface water exchanging places with the deeper water. The changes in salinity have had a major impact on the Dead Sea's sparse biota, and after the overturn, the lake seemed to be totally sterilized. Then in the rainy winter of 1979–80, the number of aquatic organisms increased dramatically, giving every indication that the Dead Sea was very much alive.

LIFE IN THE TUNDRA

One of the most barren environments on Earth is the arctic tundra of North America and Eurasia (FIG. 12-5). Tundra covers about one-tenth of the world's land surface in an irregular band winding around the top of the world, north of the tree line and south of the permanent ice sheets. Alpine tundra exists in much of the world's mountainous terrain above the tree line and below any existing glaciers. Unlike the arctic tundra, which lies at high latitudes and therefore is deprived of sunlight during the long arctic winters, alpine tundra receives daily doses of sunlight. While little snow falls in much of the arctic, alpine areas receive abundant snowfall because of their elevation.

Other than climatic differences, the two regions have much in common. Their vegetation consists mostly of stunted plants, often widely separated by bare soil or rock (FIG. 12-6). Most of the ground in the arctic tundra, called *permafrost*, is frozen year round. Only the top few inches of the soil thaws dur-

FIG. 12-4. Saguaro and other desert vegetation in the Sonora desert of Arizona.

(Photo by B. Brixner, courtesy of National Park Service)

FIG. 12-5. The arctic tundra line, North of which the ground remains frozen year round.

FIG. 12-6. Tundra terrain on Gunsight Mountain, Alaska.

ing the short summer. Therefore, runoff has nowhere to go except into countless lakes, ponds, and large bogs, where grasses and sedges often grow. However, much of the tundra is barren, mainly because of limited food resources, high winds, and frigid temperatures during most of the year. Parts of the arctic tundra are the world's most nutrient-poor habitats, and yet, algae and aquatic insects proliferate in the cold streams that eventually find their way to the Arctic Ocean.

The growing season in the arctic tundra is generally only two to three months long. Even though the ground is bathed in 24-hour sunlight, the soil temperature seldom rises much above the freezing point because most of the radiant energy goes to melting the ice in the soil. Often over a long period of time, the freeze-thaw sequence produces bizarre patterns in the ground. Some patterns are regular polygonal shapes caused by a process in which the freezing of the moisture in the soil causes the ground to heave up.

In this land of unusual ground conditions, plants and animals must make the best of limited growing season, rainfall, and nutrients. Special survival adaptations are needed in this demanding environment, which produces hardy as well as frail species. A classic example is seeds of the arctic lupine. Scientists collected these seeds from a lemming burrow that dated toward the end of the last ice age. The remarkable thing about these seeds was that they were well preserved, and when planted, they actually grew into plants that blossomed with delicate flowers. Although other ancient seeds such as wheat and corn found in Egyptian tombs and elsewhere have been grown, these are by far the oldest seeds that have germinated, thus demonstrating the hardiness of tundra species.

Despite its seeming hardiness, however, the arctic tundra is also one of the most fragile environments, and small disturbances can wreak a lot of damage. Overgrazing by reindeer on the sparse grassland can decimate large areas. Exploration activities for petroleum and minerals can ruin huge acreages. Tracks of vehicles that traveled cross-country are still visible 40 years later.

As the polar front swoops across polluted regions of the Northern Hemisphere during the winter, it transports atmospheric pollution to the arctic regions, contaminating the once pristine skies and producing a phenomenon known as *arctic haze*. The haze, which originates mostly in Europe and northwest Asia and can be as bad as that of some American suburbs, tends to block out the sunlight, which is at a premium at these high latitudes. The scarcity of large amounts of precipitation means that the pollutants are not readily scrubbed out of the atmosphere and remain aloft for long periods. When the precipitation does hit the ground, its acidity can be particularly harmful to delicate tundra flora. Lichens can exist where no other plants can and grow ever so slowly. Because they absorb dangerous substances from atmospheric moisture and dust which can build to lethal levels, they are usually among the first to perish. Other animals feeding on the contaminated lichens accumulate the poisons in their bodies, which could prove to be an ecological disaster.

LIFE UNDER THE ICE

Sea ice in the Arctic Ocean (FIG. 12-7) inhibits the growth of *phytoplankton* for two-thirds of the year. In this environment, algae become trapped in the sea ice at freeze up. During the long, dark winter, their growth is very slow, but as the light returns in the early spring, they proliferate rapidly at the ice-water interface.

In such cold, dark environments, many species of herbivorous zooplankton like the tiny plant-eating crustaceans called copepods, are an important link in the marine food chain and require two or more years to complete their life cycles. Reproduction usually takes place just before or immediately after the breakup of the ice, and produces a single brood. Infant development is relatively rapid during the open-water season from April to September, with most individuals reaching adolescence before the onset of new ice. During the winter, from October until about mid-May, there is virtually no phytoplankton in the seawater, and yet, in the late winter or early

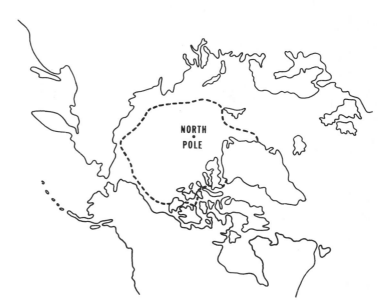

FIG. 12-7. Extent of Arctic ice in September.

NORTH POLE

spring, the young copepods somehow emerge as mature adults. Apparently, the hungry copepods overwinter beneath the ice, grazing off the ice algae. It is not known whether they can actually remove algae from under the ice or simply consume the algae melted at the ice-water interface. Nevertheless, they survive the ordeals of the harsh winter and continue to make the polar oceans one of the richest environments on Earth.

The arctic accounts for some 10 percent of the world's fish catch; among the most important are cod, capelin, and herring. The Bering Sea is the world's most productive region for walleye and pollock. The pollock alone has an estimated yearly yield of over 1 million tons, accounting for nearly 80 percent of the foreign catch in the Bering Sea.

In Alaska, the salmon fishery is particularly important, especially to the natives. In the North Sea, the total fish catch is about 3 million tons annually. Although the amount has remained fairly steady since the late 1960s, there has been a pronounced decline in the catch of heavily exploited herring and mackerel. These losses have been compensated mainly by increased yields of cod, haddock, pollock, and whiting, along with other small fishes that are generally considered trash fish.

It is not clear to what extent these changes are a result of shifts in fish populations, changes in patterns of commercial fishing, or environmental stress. Population changes tend to be more variable and unpredictable because of the strongly seasonal behavior of the fishes and the significant differences in climate and other environmental conditions from one season to the next. Arctic marine mammals, especially seals, are also considered commercially valuable. However, some mammals such as the bowhead whale and some fish such as the arctic cisco have been so overly exploited they are now endangered species.

Part of this abundance in the polar regions is the result of the ability of some fish species to survive in subfreezing seawater. The Antarctic Ocean is the coldest marine habitat in the world. It is cut off from the general circulation of the ocean by a circum-Antarctic Current (FIG. 12-8), which produces a thermal barrier that impedes the inflow of warm currents as well as the outflow of antarctic fishes. Four months out of the year, Antarctica is in total darkness, and sea ice covers the water for at least ten months of the year (FIG. 12-9). Even in the short summer, the water under the ice receives less than 1 percent of the sunlight on the surface.

Life Under the Ice 163

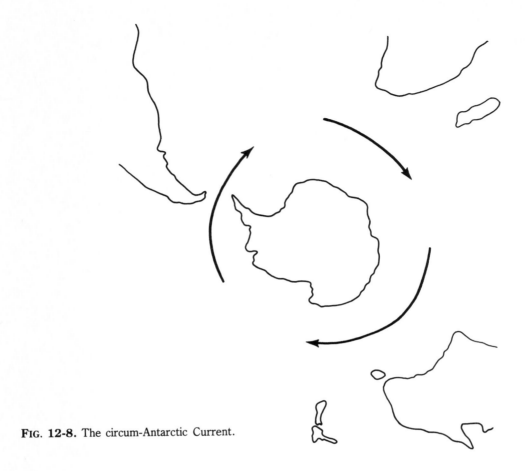

FIG. 12-8. The circum-Antarctic Current.

LIFE IN HOT WATER

It is because of these extreme conditions that the Antarctic Ocean has only about half the diversity of its northern counterpart. Yet, despite the fact that seawater remains a few degrees below freezing throughout the year, certain species of fish thrive beneath the Antarctic ice. The salt content and other substances in the blood of these fish act as a sort of antifreeze to prevent the propagation of ice crystals in their bodies. Because fish are cold-blooded, their body temperature is essentially the same as their environment. Therefore, in order to survive the long, dark antarctic winters when the food supply is scarce, fish must conserve energy using neutral buoyancy, as well as prevent freeze up.

Organisms that are capable of living at high temperatures are called *thermophiles* (heat-lovers) and have long fascinated biologists and earth scientists alike. Natural high-temperature environments are widely distributed on the Earth and are generally found in association with volcanic activity (FIG. 12-10).

No multicellular plants and animals can live for very long in water at temperatures above 50 degrees Celsius (120 degrees Fahrenheit); thus, only microorganisms are found above this temperature. Among the microorganisms, the upper temperature limit seems to be about 60 degrees Celsius (140

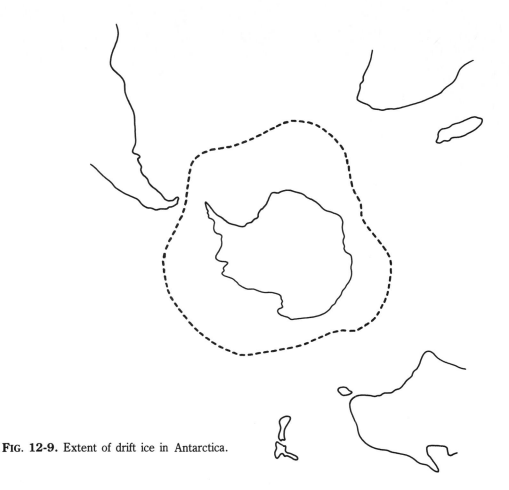

FIG. 12-9. Extent of drift ice in Antarctica.

degrees Fahrenheit), for eukaryotes probably because their nucleus is unable to function at higher temperatures. Procaryotes, which lack a nucleus like certain bacteria, are found in most boiling-water environments where they often reproduce extremely well.

Apparently, bacteria are able to grow at any temperature at which water remains a liquid, even in pools that are above the normal boiling point of water. The reason is that the temperature at which water boils is determined by the pressure, and although water boils at 100 degrees Celsius (212 degrees Fahrenheit) at sea level, water can still re-main a liquid at several times this temperature at the ocean depths. Biological constraints place the upper temperature limit at about 200 degrees Celsius (390 degrees Fahrenheit), above which amino acids are destroyed. Therefore, the upper temperature limit for life might be anywhere between 100 and 200 degrees Celsius.

The origin of thermophilic microorganisms has intrigued scientists for over 100 years, and two opposing hypotheses have been put forward to explain their existence. If the first organisms arose in high-temperature environments, thermophilic organisms would have been primordial, and all subsequent or-

Fig. 12-10. Hot water river flowing in Yellowstone National Park, Wyoming.

(Photo by K.E. Bargar, courtesy USGS)

ganisms would have derived from them. On the other hand, if the first organisms were not thermophiles, then thermophilic organisms might have had a secondary origin from colder temperature species called *mesophiles*. However, many genetic changes would have had to take place for a thermophilic organism to be derived from a mesophile, whereas, if the first organisms were thermophiles, it would be much easier for a mutant to evolve that was incapable of growing at higher temperatures.

Thus, the common ancestral organism of all life forms was probably a thermophile. The thermophiles frequently have a sulfur-based energy metabolism, and sulfur compounds would have been readily available on the hot, infant Earth.

LIFE IN THE ABYSS

It was once thought that the deep waters of the open ocean were a lifeless desert. In the 1870s, the British oceanography ship H.M.S. *Challenger* dredged the bottom of the Pacific Ocean in some of the deepest parts. It brought to the surface a collection of bizarre creatures, some species of which were thought to be long extinct. More recently, a population of large, active animals were found thriving in total darkness as much as four miles deep. This environment was once thought to be the domain of small feeble creatures, such as sponges, worms, and snails, that were specially adapted to live off the debris of dead animals raining down from above. Much of the deep-sea floor was found to be teeming with numerous species of scavengers, including highly aggressive worms, large crustaceans, giant squids, deep-diving octopuses, a variety of fishes, and giant sharks.

Part of the reason for the large size of many species at this depth is an abundance of food, lower levels of competition, and lack of juveniles, which spend their time in shallower depths and descend to deeper water when they mature. Large numbers of fish existing at great depths in the lower latitudes are related to shallow-water varieties living in the higher latitudes. Some species of arctic fishes apparently represent near-surface expressions of populations that inhabit the cold, deep waters off continental margins to the south.

One of the strangest environments on Earth is located in deep water near seafloor-spreading centers such as the crests of the East Pacific Rise and the Mid-Atlantic Ridge. At the bases of jagged basalt cliffs is found evidence of active lava flows and fields strewn with pillow lava. The pillow lava forms when molten rock is ejected out on the ocean floor and is cooled quickly by the cold water.

In this area of the ocean, exotic looking chimneys called *black smokers* spew out hot water blackened with sulfide minerals. Others eject hot water that is milky white and are called *white smokers*. The hot water comes from deep below the surface where seawater percolating through cracks in the ocean crust is heated near magma chambers below the spreading centers. The seawater rises up to the surface and is expelled through hydrothermal vents like undersea geysers (FIG. 12-11).

In these volcanically active fields is a habitat unlike any other in the world. The hydrothermal vents not only keep the bottom waters at a reasonable temperature—upwards of 70 degrees Fahrenheit—they also provide valuable nutrients, making this the only environment that is totally independent of the Sun as a source of energy. Instead, the energy comes from the Earth's interior.

In 1977, while exploring the East Pacific Rise off the tip of Baja California, the research submarine *Alvin* discovered an oasis in water 8000 feet deep. Species previously unknown to science were living there in total darkness. Tube worms three feet tall swayed eerily in the hydrothermal currents. Giant white crabs scampered blindly across the volcanic terrain. Large white clams up to a foot long and clusters of mussels formed large communities around the hydrothermal vents.

The most remarkable thing about these animals is that they do not receive their food supply in the form of detrital material falling from above, but instead rely on symbiosis with a certain sulfur-eating bacteria that live within their tissues. The bacteria harnessed energy liberated by the oxidation of hydrogen sulfide from the vents in order to incor-

porate carbon dioxide in the production of organic compounds such as carbohydrates, proteins, and lipids. The by-products of the bacteria's metabolism leaked into the host animal and nourished it.

The vent animals are so dependent on the sulfide-metabolizing bacteria that the mussels have only a rudimentary stomach and the tube worms lack even a mouth. Some of the animals also feed on the bacteria directly. All the animals live precarious lives because the hydrothermal vent systems turn on and off almost at will, and the animals can continue to survive only as long as the vents continue to operate.

Biologists of the Smithsonian Institution using a deep-sea submersible made a remarkable discovery off the Bahamas in 1983, which could open up a whole new realm of oceanography. On an uncharted seamount, the scientists found a totally new and unexpected plant at a depth of about 900 feet—deeper than any previously known oceanic plant larger than a microbe. It had been generally thought that the lowest depth plants can survive was around 600 feet because very little sunlight penetrates below that level. The new plant is a variety of purple algae with a unique structure, consisting of heavily calcified lateral walls but very thin upper and lower walls that enable the plant to make the most of the

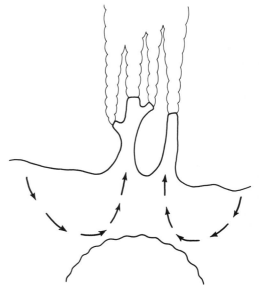

FIG. 12-11. Operation of hydrothermal vents.

scant amount of sunlight that reaches this region of the ocean. The cells are stacked on top of each other like cans at a grocery store for maximum surface exposure to the feeble light.

The discovery of this unusual plant will force scientists to rethink the role that such algae play in the productivity of the oceans, marine food chains, sedimentary processes, and reef building. It is also one more clear example that the deep-sea floor still remains for the most part virgin territory, and exemplifies the fact that the exploration of inner space is just as important as the exploration of outer space.

13
Extinctions Are a Way Of Life

PRACTICALLY all plant and animal species that have ever existed on the face of the Earth are extinct. Through eons, species have come and gone so that those presently living represent only a small fraction of all those that have gone before. It is for this reason that extinctions play an enormous role in the evolution of life. When there is a major extinction, new species develop to fill vacated niches.

During the Phanerozoic eon, or the past 600 million years, there have been five major mass extinctions and five minor ones. All seem to indicate biological systems in extreme stress brought on by a drop in sea level or climatic change.

There is some speculation that extinctions are periodic, brought on by celestial causes such as cosmic rays from supernovas or meteor impacts. There is even a comparison made between these catastrophic events and nuclear war, whose environmental impact is thought to be much the same. Extinctions also might be episodic, with relatively long periods of stability followed by seemingly random, short-lived (geologically speaking) extinction events. Most extinction episodes also select certain types of species, and analysis of the victims and the survivors might help resolve the main causes of the extinctions.

The extinctions of the past, whether they were periodic or episodic, were caused by natural phenomena. Present-day extinctions, however, in which something on the order of 100 species vanish every day, are *forced extinctions* caused by man's activities. Once a species is gone it is gone forever. These extinctions have been justified by saying that man is also a part of the natural world.

If the present rate of extinctions continues to climb geometrically just as man's population and environmental destruction continues to climb geometrically, 50 percent or more of the species now living will disappear sometime during the next century, making this one of the worst extinction events of the entire history of the Earth.

THE EXTINCTION CYCLE

As this book has shown, the geologic time scale is divided into eras, which are further subdivided into

Table 13-1. Radiation and Extinction for Major Organisms.

ORGANISM	RADIATION	EXTINCTION
Marine invertebrates	Lower Paleozoic	Permian
Foraminiferans	Silurian	Permian & Triassic
Graptolites	Ordovician	Silurian & Devonian
Brachiopods	Ordovician	Devonian & Carboniferous
Nautiloids	Ordovician	Mississippian
Ammonoids	Devonian	Upper Cretaceous
Trilobites	Cambrian	Carboniferous & Permian
Crinoids	Ordovician	Upper Permian
Fishes	Devonian	Pennsylvanian
Land plants	Devonian	Permian
Insects	Upper Paleozoic	
Amphibians	Pennsylvanian	Permian-Triassic
Reptiles	Permian	Upper Cretaceous
Mammals	Paleocene	Pleistocene

periods. The boundaries of the periods represent major extinctions of species. The relationship between one period and the next is based on relative time, which simply means that one is older than the other according to fossil evidence. When attempts were made to attach absolute time units based on nuclear decay to the relative time scale, difficulties arose that were only partly resolved by making the best fit possible between two very different time scales.

Over the last 250 million years, there have been eight major extinction events. Some of the strongest peaks coincided with the boundaries between major geological periods. What is more, the episodes appeared to be periodic, occurring around every 30 million years. Moreover, longer period cycles of 80 to 90 million years appear between major mass extinctions, with exceptionally strong mass extinctions occurring every 225 to 275 million years.

The long-period mass extinctions might be related to the period of the Sun's revolution around the center of the Galaxy, which takes 200 to 250 million years. Should the Solar System enter a nebula or a dust cloud on its journey around the Galaxy, (FIG. 13-1) the diffused dust could affect the Sun's output or block solar energy to the Earth and cooling, which would have an effect on the biosphere. The argument against periodicity asserts that major extinctions might be just a clustering of several minor events at certain times that only seem to be cyclical. In other words, random groupings of extinct species on a geologic time scale that itself is uncertain might be nothing more than coincidence.

Astronomical phenomena might best explain the apparent periodicity of the extinctions. The Sun, like all the stars in the Galaxy, oscillates up and down perpendicular to the galactic plane, completing a cycle about every 60 to 80 million years—which is close to one of the major extinction cycles and about twice the 30 million-year cycle of extinction. The Sun, therefore, crosses the midplane and reaches the maximum distance from the midplane twice each cycle, or roughly every 30 million years. Furthermore, the amplitude of the oscillation is calculated to be approximately 100 parsecs (1 parsec is 3.26 light-years) on either side of the midplane of the Galaxy.

The passage of the Sun through dense clouds located at the midplane could reduce the Earth's solar irradiation and initiate climatic changes that could dramatically affect life on the planet. There is, however, no evidence that the dust cloud is sufficiently thick to block out the Sun during each passage through the midplane, which can take several million years to complete.

Presently, the Sun is near the midplane of the Galaxy, and we appear to be midway between major extinction events, of which the most recent ones were approximately 11 million and 38 million years ago. The extinction episodes therefore might instead coincide with the approach of the Sun to its extreme positions away from the galactic plane. The extinction of the dinosaurs and large numbers of other species 65 million years ago occurred at a time when the Sun's distance from the midplane was near its maximum. As the Sun reaches the upper or lower regions of the Galaxy, it might become more vulnerable to higher levels of cosmic radiation from an exploding supernova, which could ionize the Earth's upper atmosphere and produce a haze that would block out sunlight.

About twice every million years, the Earth's magnetic field reverses itself. After a long stable period of hundreds of thousands of years, the strength of the magnetic field gradually decays over a short period of several thousand years. At some point, the field collapses all together and after some time, is regenerated in the opposite direction. During the collapse, the Earth is totally without a magnetic field, and the magnetosphere disappears. It is believed that the Earth's core operates similarly to a disk dynamo. In the laboratory, the magnetic field generated by the dynamo randomly collapses and is regenerated with opposite poles. Some scientists don't believe this collapse is a purely random event, but maintain it can be described mathematically.

The magnetosphere shields the Earth from particle radiation in the solar wind (FIG. 13-2), which is responsible for the northern and southern lights seen at night in the higher latitudes. The bombardment of radiation in the solar wind could influence the composition of the upper atmosphere by producing higher levels of nitrogen oxides, which in turn could produce a haze that blocks out the Sun. A comparison between magnetic reversals and variations in the climate has shown in many cases a striking agreement.

Certain magnetic reversals also coincide with the extinction of species. Nevertheless, it has not been convincingly demonstrated that magnetic reversals are periodic in themselves, although they might be associated with other periodic phenomena. One suggestion is that the magnetic field reversals might be triggered by variations in the galactic magnetic field as the Sun moves through the galactic plane. The galactic magnetic field is so weak, however, that it is doubtful whether it would have any influence on the Earth's magnetic field, which is a million times stronger. Large meteor impacts, very strong earthquakes, or intense volcanic activity also have been cited as causes of geomagnetic field reversals.

One explanation for the extinctions that has

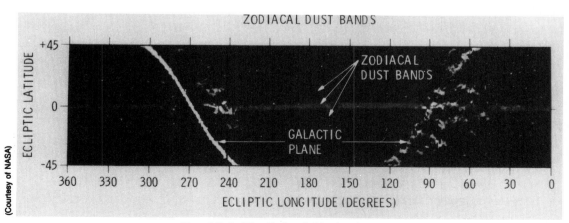

FIG. 13-1. The Zodiacal dust bands.

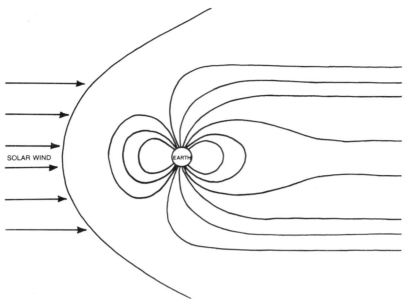

FIG. **13-2.** The Earth's magnetic field shields the planet from harmful radiation in the solar wind.

SOLAR WIND

EARTH

caught much attention lately is the *impact theory*. The boundary rocks between the Cretaceous (abbreviated K) and the Tertiary (T) periods, known as the K/T boundary contain, a layer of mud composed of shock-impact sediments, an unusually high irridium content, and a thin layer of organic carbon. Irridium is an isotope of platinum, which is rare on Earth but is relatively abundant in asteroids and comets. An exceptionally large meteor striking the Earth with a force greater than all the nuclear arsenals in the world would send aloft large amounts of sediments into the atmosphere, along with soot from massive forest fires that would be set ablaze by impact friction. The debris could clog the skies and place the planet in a deep freeze, a scenario used to describe "nuclear winter." It is argued that a number of large volcanic eruptions could produce the same phenomenon.

Going back through the geologic record, scientists have uncovered more layers having anomalous irridium concentrations. These layers coincide with other extinction episodes. The periodicity of the extinctions, which appears to be about every 26 million years, might be attributed to Earth's movement through the galactic plane whose gravity disturbance

might break loose comets in the Oort Cloud. It has also been suggested that the Sun has a companion brown dwarf star named Nemesis that passes close by the Oort Cloud every 26 million years, thus disturbing a large number of comets which then shower down on the Earth.

Cycles of mass extinction also correlate with cycles of terrestrial phenomena. The largest of these cycles lasted about 300 million years and is related to the cycle of convection currents in the Earth's mantle. It began with rapid convection in the mantle, which led to the breakup of supercontinents. This breakup squeezed the ocean basins, causing a rise in sea level and transgressions of the seas inland. It also increased volcanism, which in turn increased atmospheric carbon dioxide. The result was a greenhouse effect that produced warm conditions worldwide. These episodes occurred from the Late Cambrian to the Late Devonian periods (roughly from 500 to 350 million years ago) and from the Early Jurassic period to the Late Eocene epoch (roughly from 200 to 50 million years ago).

The second phase of the 300-million year cycle was a time of low mantle convection, the assembly of continents into a supercontinent, the widening of ocean basins with a consequent drop in sea level and a regression of the sea from the land, a reduction of atmospheric carbon dioxide by low levels of volcanism, and the development of an icehouse effect with colder temperatures worldwide. These conditions prevailed from the Late Precambrian to the Early Cambrian periods (roughly from 700 to 550 million years ago), from the Late Paleozoic era to the Triassic period (roughly from 400 to 250 million years ago), and during the latter part of the Cenozoic era (the period in which we are now living).

TERRESTRIAL CAUSES OF EXTINCTION

The drifting of the continents has a dramatic effect on the Earth's biota. Continents and ocean basins are continually being reshaped and rearranged by lithologic plates, which are constantly in motion. When continents break up, they tend to override ocean basins, making the seas less confined and thus raising the sea level several hundred feet. Low-lying areas inland of continents are inundated with seas, which dramatically increases the amount of shoreline and therefore the amount of shallow-water marine habitat. Mountain building associated with the movement of plates alters patterns of river drainages and climate, which affects terrestrial habitats. The placement of continents in separate parts of the world also can interfere with ocean currents, which affect the distribution of the Earth's heat.

The opposite condition occurred when continents were assembled into a supercontinent during the latter part of the Paleozoic. Freeflowing ocean currents distributed heat from the tropics to the poles and kept the temperature of the planet more uniform. The ocean basins widened, which caused the sea level to drop considerably, forcing the inland seas to retreat and producing a continuous, narrow continental margin around the supercontinent. The amount of shoreline was greatly reduced, which radically reduced the habitat area. Unstable nearshore conditions resulted in an unstable food supply. Species that were elaborately adapted specialists existing in a variety of ecosystems could not cope with the limited living space and limited food supply. As a result, 50 percent of the families (FIG.

13-3) and 95 percent of all marine species became extinct at the end of the Paleozoic, 240 million years ago—the greatest mass extinction the Earth has ever endured.

Some mass extinctions also coincide with periods of glaciation because temperature is perhaps the single most important factor limiting the geographical distribution of species. Certain species, like coral, can survive only within a narrow range of temperatures (FIG. 13-4). During warm interglacial periods, species invade all latitudes, but as glaciers advance across the continents and ocean temperatures drop, species are forced into warmer regions. Species that are unable to adapt to the colder conditions or that are attached to the ocean floor and are unable to migrate to warmer areas are usually the ones hit hardest. The intense competition for habitat and food severely limits diversification and thus the number of species. The locking up of great amounts of seawater into glacial ice significantly lowers the sea level, which also limits shallow-water habitat and causes crowding conditions worldwide.

Not all climatic cooling is accompanied by glaciation, which is dependent on many factors, including the tilt of the Earth's axis, the position of the continents, and the circulation of the oceanic and atmospheric currents. The greatest glaciation took place toward the latter part of the Precambrian, about 680 million years ago. Phytoplankton, which were the highest form of animal life at that time, were decimated in the cold waters of the ocean. When the ice disappeared near the end of the era, there was an explosion of species as the oceans began to warm again.

A large number of volcanic eruptions operating over a long period could lower global temperatures substantially by injecting huge amounts of volcanic ash and dust into the upper atmosphere, thereby blocking the Sun's energy and causing mass extinctions of plants and animals. Heavy clouds of volcanic dust have a high albedo and reflect much of the solar radiation back into space. Some of the radiation goes to warm the dust itself and some is scattered sideways and still reaches the ground, but at an indirect angle. The effect is to shade the Earth and

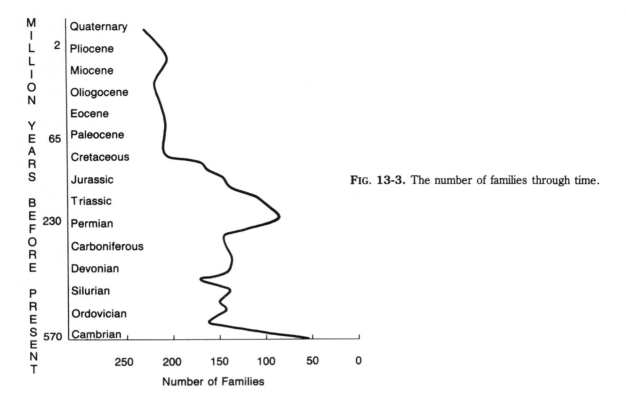

FIG. 13-3. The number of families through time.

lower the temperature. Even a 5 percent drop in solar radiation could cause the surface to cool by as much as 10 degrees Fahrenheit which would be enough to bring on an ice age.

The climatic effects of volcanic dust are dependent on the type and size of the particles being released and where they are located in the atmosphere. The smaller particles, the *aerosols*, in the stratosphere tend to remain the longest and reflect sunlight the most, while allowing surface heat to escape into space. The long-term cooling could cause glaciers to expand, thus lowering the sea level and limiting the habitat area. The lowered temperature also could adversely affect the distribution of species and confine warmth-loving species to regions in the tropics.

Some geologists argue that the pressure-shocked crystals and the abundance of irridium in the thin sediments between rocks of the Cretaceous and Triassic periods are not necessarily from me-

teor impacts. Instead, they could have been produced by powerful volcanic eruptions.

Large meteor impacts (FIG. 13-5) would have the same environmental consequences as large volcanic eruptions. A single massive meteor or several moderate-sized ones could gouge out enough earth material and send it into the atmosphere to cause darkness at noon for up to six months or longer. Photosynthesis would be stopped and near-surface phytoplankton in the ocean would be eliminated. The effects of these extinctions would cascade up the food chain in a sort of domino effect, killing off large marine, as well as large terrestrial, species.

Such an extinction event would appear to be instantaneous in the geologic record. However, a resolution of several thousand years, much less several years, over millions of years of geologic history is not possible. It appears more likely that the extinction of species occurred over lengthy periods, perhaps 1 million years or more, and because of ero-

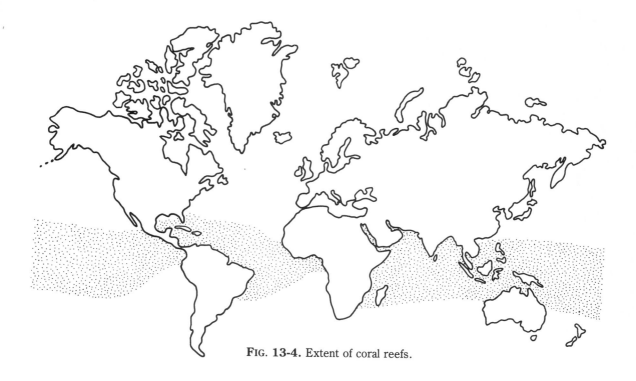

FIG. 13-4. Extent of coral reefs.

FIG. 13-5. Aerial view of Meteor Crater, Coconino County, Arizona.

(Photo by W.B. Hamilton, courtesy of USGS)

sion or nondeposition, the die-outs only appear to be sudden.

During the last part of the Cretaceous, the ammonites (FIG. 13-6), which were a highly success-ful species, completely died out during an interval of about 2 million years, probably because of the development of more mobile predators, although it is highly unusual in nature for predators to totally decimate their prey, since they would themselves become decimated. What is more, the ammonites disappeared some 100,000 years before the supposed meteor impact. Other species, including the dinosaurs (FIG. 13-7), were already in decline several million years prior to the end of the Cretaceous. It seems then that only a massive meteor shower spread out over a similar length of time would show such a lengthy period of mass extinction that affected the biota in all parts of the Earth.

EFFECTS OF EXTINCTIONS

Throughout the history of life on Earth there have been gradual extinctions, called *background extinctions*. Major extinction events are separated by periods of lower extinction rates, and the difference between the two is only a matter of degree. The relationship can be compared to war, in which there are long periods of boredom punctuated by short periods of terror.

There is a qualitative, as well as a quantitative, distinction between background and major extinctions. Species have regularly come and gone even during optimum conditions. It was likely that they lost their competitive edge and were nudged out by a superior, better adapted species. However, just because a species survives extinction does not necessarily mean it is better in these respects, but that the losers were developing certain unfavorable traits such as elaborate physical and behavioral patterns during background times. Therefore, it appears that those characteristics which permit a species to live successfully during normal periods

FIG. 13-6. The ammonites became extinct at the end of the Cretaceous.

FIG. **13-7.** Triceratops was one of the last dinosaurs to become extinct.

become irrelevant when major extinction events occur. Also, the distinction between background and major extinctions might be clouded by ambiguities in the fossil record, particularly when certain species are favored over others for fossilization.

Each time there is a mass extinction, it resets the evolutionary clock, as though life were forced to start anew. Those species that survive radiated outward to fill entirely new niches, which in turn, produce entirely new species. A separate species is defined more by genetics than by morphology. Different species might share certain similar physical attributes because they share the same niches. Because they have different genes, however, crossbreeding among species is impossible. The new species might develop novel adaptations that give them a survival advantage over other species. The adaptations might lead to exotic-looking species that prosper during normal background extinctions but because of their overspecialization, are totally incapable of surviving mass extinctions. Thus, the fossil record shows a myriad of strange species (FIG. 13-8), the likes of which have never been seen since.

Nature is constantly experimenting with different life forms, and when one becomes a failure, it is relegated to the trash heap of extinction, never to be tried again. Once a species is gone, it is gone forever, and the odds that its particular combination of genes will reappear are astronomical. As a result, evolution is on a one-way street, and although species are perfected to live their optimum in their respective environments, they can never go back to the past. That is why even though today's environment might perfectly match that of the trilobites (FIG. 13-9), they will never return—neither will the dinosaurs.

The mammals were able to replace the dinosaurs, not because they were more intelligent or because they were warm blooded, which are adaptations that would seem to give them a decisive edge during times of environmental stress. It probably had more to do with their small size, al-

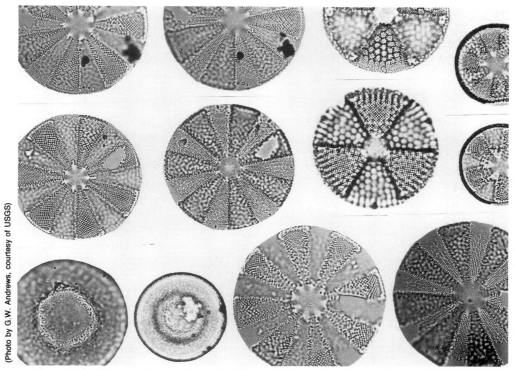

(Photo by G.W. Andrews, courtesy of USGS)

FIG. 13-8. Miocene diatoms from the Choptank Formation, Calvert County, Maryland.

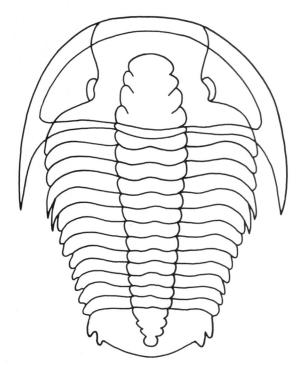

FIG. 13-9. Trilobites first appeared in abundance in the Cambrian period and became extinct at the close of the Paleozoic era.

though not all dinosaurs were giants and many were no larger than most mammals today. The mammals also might have preyed on dinosaur eggs and won the battle of evolution through attrition.

Whatever were the mammals' advantages, they were able to outcompete and outpopulate the dinosaurs, and over a period of several million years, they were able to replace them completely. The fact that the dinosaurs were not the only ones to go and that 70 percent of all known species became extinct at the end of the Cretaceous period indicates that something in the environment made them all unfit to survive.

The dinosaurs and the other species could have been wiped out by intense acid rain caused by a huge meteorite impact 65 million years ago. The downpour of highly corrosive acid could have destroyed plants and the animals that fed on them. Global temperatures could have been raised by the disappear-

ance of marine algae, which are a major source of dimethyl sulfate molecules that act as cloud-making nuclei. Without clouds to reflect the Sun's rays back into space, the Earth could have heated up substantially, making conditions intolerable for life.

Whatever the reason, it did not seem to adversely affect the mammals to any large extent. It appears that the placental mammals came through the extinction and flourished at the expense of the marsupial mammals, which seems to indicate that the more advanced and superior form of placental reproduction was the reason for their success. Yet, egg-laying fish, amphibians, reptiles, and birds still enjoy a high degree of reproductive avarice, as indicated by large populations of these animals just about everywhere.

The next step is trying to explain the success of humans as a species. If environmental stress produces a new species, then humans were a byproduct of the ice ages. The factors that contributed to human success were not the same as those that favored the mammals over the dinosaurs, because the human has one quality that no other animal possesses: the conscientious ability to change his environment to suit his purpose, rather than just being a passive passenger on the evolutionary train. As long as humans remained primitive and their populations remained within the carrying capacity of their natural environment, they were of no great threat to other species, and probably were a beneficial force in keeping the population of certain species in balance.

It was only after the end of the last ice age, when the human spirit is said to have soared along with its population, that the pressure of human presence was felt by the rest of the living world. It is argued convincingly that certain species of large mammals, like the mastodon and mammoth (FIG. 13-10), became extinct toward the end of that ice age, not from competition from a more successful species or by environmental stress—which are the normal channels by which species become extinct—but instead from decimation by the hand of man in disregard of the balancing effect of the predator-prey relationship. It is in this respect that man appears to be outside the control of nature.

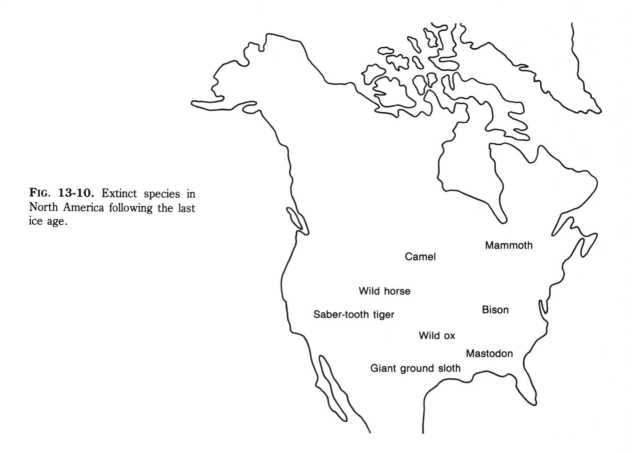

FIG. 13-10. Extinct species in North America following the last ice age.

Mammoth

Camel

Wild horse

Saber-tooth tiger

Bison

Wild ox

Mastodon

Giant ground sloth

MODERN EXTINCTIONS

Scientists trying to understand the disappearance of the dinosaurs and other species at the end of the Cretaceous have come up with some exotic explanations, including everything from meteor impacts to exploding supernovas. It is much easier to place a finger on the reasons for modern-day extinctions, which if compared with those of the past, will not take millions nor thousands of years, but instead only a century. Large marine mammals such as the whales and manatees appear to be going the same direction as their reptilian counterparts, but the extinction pressure is not coming from climatic changes or by some fierce marine predator, but instead from man.

Despite efforts to stop the extinction of certain species, such as halting the construction of dams to save darter snails, the number of animals that are disappearing is greatly exceeding the natural background rate and is estimated to be on the order of about 100 species per day. If the present spiral of human population and environmental destruction continues out of control, half or more of the species living on Earth today will be gone by sometime in the next century. In effect, they will have been destroyed by a sort of human volcano that is immensely more destructive than anything nature has managed to produce. Future geologists millions of years from now will view this period as a very sharp and powerful spike in the fossil record—possibly the greatest die-out of species in the shortest time in geologic history.

Man is, therefore, in the unique position of being the only animal that has ever existed that has

FIG. 13-11. The world environments. A = tundra, B = forest, C = savannah, D = desert.

effected the extinctions of large numbers of other species. Even the dinosaurs, which were perhaps the most successful creatures on Earth, could not do that. All animals that are not beneficial to man, and even some that are, will be forced aside as growing human populations continue to squander the Earth's space and resources and contaminate the soil, water, and air.

The complex interrelationships between species and between them and their environments (FIG. 13-11) is still not fully understood, but it is probably safe to assume that the destruction of large numbers of different species will not just leave the world devoid of beneficial animals, but will allow those species commonly considered as pests and parasites to flourish because their natural predators will be gone. Therefore, not only will we lose much of the great beauty and wonder of the Earth, but we will gain some species that will not make this the best of all possible worlds.

Also, it is not known exactly how many species there actually are. Many are hidden out of sight, so their benefit to man goes largely unnoticed. In addition, it is not just the larger animals that are dying off, but also simple creatures at the beginning of the food chain: photosynthesizors that produce oxygen, and microbes that aid in nitrogen fixation for plants. If they go, we go. The extinction of our own species is not an easy thing to contemplate, nor does it require anything as catastrophic as nuclear war. If present trends of environmental destruction continue, it will be all that is necessary—it will just take a little longer.

14

Life Hanging
in the Balance

Humans are irreversibly changing the face of the Earth by substantially altering the physical as well as the biological environment. In 1987, the population reached 5 billion people, and there are predictions that the Earth's carrying capacity can handle several times that many. Unfortunately, in order to feed a large population of people, the rest of the world will have to move aside to make room for additional agriculture and urbanization.

The tropical forests are being laid waste at an alarming scale for agricultural land. Along with the forests goes millions of species of plants and animals. Through improper farming techniques, topsoil is eroding away at several times the replacement rate. Man-made deserts are spreading throughout the world, and even natural deserts are encroaching on agricultural lands. Large amounts of carbon dioxide from the combustion of fossil fuels and the destruction of forests are being sent into the atmosphere, which could raise global temperatures and adversely affect weather patterns, making areas either much too dry or much too wet. Combustion of fossil fuels also produces deadly acid rain, which is destroying

forests, crops, fish, and much of the ancient beauty handed down from earlier cultures. Pollutants are slowly eating away the ozone layer in the upper atmosphere, causing a higher incidence of skin cancer and a loss of crops and aquatic life. Overfishing in the oceans is destroying good varieties of fish and allowing unwanted species to flourish. Marine species are also being poisoned by water pollution and dumping of wastes into the ocean. If we allow these problems to go unsolved, they will most likely solve themselves—but the outcome will not be a pleasant one.

THE CLOUDED SKIES

Much of the air pollution that reduces visibility and harms plants and animals, as well as man-made structures, is composed of *dry deposits*. These atmospheric particles consist of unburned carbon, dust particles, and minute sulfate particles. The finest particles come mainly from chemical processes found in the combustion of fossil fuels, and are essentially acidic. Because of their small size, these particles, called *aerosols*, scatter light that normally heats the ground so that it heats the atmosphere and causes a temperature imbalance between the atmosphere and the Earth's surface. The results are abnormal weather patterns in all parts of the world.

High-temperature combustion yields nitrogen oxide gases, along with gaseous nitric acid. Photochemical smog in the big cities comes from the reaction of sunlight on these chemicals, which initiates a chain of complex reactions. The Great Smoky Mountains of eastern Tennessee once produced a natural photochemical haze, hence their name. Today, it seems that much of the air pollution in the mountains is a byproduct of distant coal-fired plants.

The coarse particles in the air are derived mainly from the mechanical breakup of naturally occurring materials, such as those produced by volcanic eruptions and dust storms. Large particles of carbon, called *soot*, are produced from forest fires and inefficient combustion of fuels such as in wood-burning stoves. Under certain atmospheric conditions, these particles can produce a persistent haze during the winter. Even in pristine areas where the air is considered clean, significant levels of sulfate-containing particles are found, indicating that they came from distant sources.

In an effort to eliminate local air-pollution problems from coal-fired plants, smokestacks were built tall so that pollutants would mix with the turbulent air above and be carried aloft. However, this solution also created regional air-pollution problems by allowing pollutants to be carried for long distances, causing a host of political problems downwind. Sweden and Norway are constantly being bombarded with pollution from the heavily industrialized regions of Great Britain and Germany. As mentioned in Chapter 12, even the arctic tundra is not immune and is invaded by air pollution, especially in the wintertime.

The increase in atmospheric carbon dioxide levels (FIG. 14-1) since the beginning of the Indus-

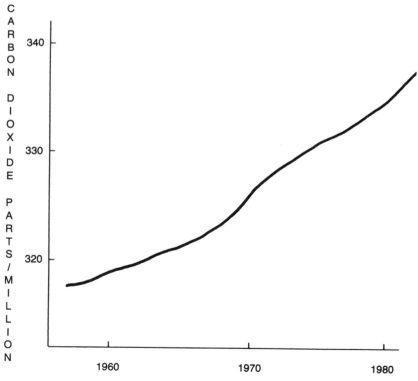

FIG. 14-1. The increase in atmospheric carbon dioxide.

trial Age has contributed to the greenhouse effect, and it has even been suggested that this increase could bring about a greening of the Earth since plants also require carbon dioxide and give off oxygen. The lush Cretaceous forests and swamps where dinosaurs ruled in all parts of the world might have been in part a result of the warm temperatures and high concentrations of atmospheric carbon dioxide. However, during that time, there were no ice caps and the seas invaded much of the land. Should the ice caps melt today, the sea level could rise some 300 feet and inundate valuable real estate. This situation is presently occurring in Venice, Italy, where subsidence from overuse of groundwater and rising sea levels from melting glaciers has dropped the city more than 8 inches relative to the sea. Warmer temperatures also could change worldwide precipitation patterns, enlarging the deserts and adversely affecting agriculture.

Most of the increase in the levels of carbon dioxide is a result of the burning of huge amounts of fossil fuels and the large-scale destruction of forests and wetlands to make way for agriculture. Moreover, croplands do not store as much carbon dioxide as the forests they replace, so less carbon dioxide is taken out of the atmosphere. Also, the oceans absorb atmospheric carbon dioxide only at a slow, steady rate, which is about half the rate it is being released into the atmosphere by man's activities.

Other serious pollutants arising from our modern life-style are the halocarbons. They are used in the manufacture of plastic foams, as refrigerants, as industrial solvents, and as spray-can propellants outside the United States. (The United States banned their use because of their link to ozone destruction.) The ozone layer in the upper stratosphere protects the Earth from deadly ultraviolet radiation (FIG. 14-2), and although small doses of ultraviolet aid the body in the manufacture of vitamin D, large amounts can lead to serious sunburn and skin cancer. Increased ultraviolet exposures also reduce crop productivity and aquatic life, especially primary producers.

During a seven-year period beginning in 1980,

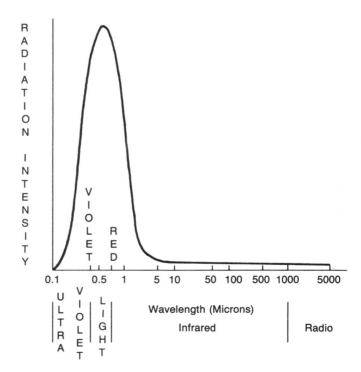

FIG. 14-2. The solar spectrum.

there was an observable drop in atmospheric ozone of about 4 percent. The ozone hole over the Antarctic, as well as another one discovered over the Arctic, are strongly believed to have a chemical origin and that the chemicals are man-made. The amount of chlorine monoxide, which destroys ozone molecules, was found to be 100 times the normal amount in the Antarctic stratosphere.

In addition to halocarbons, some 200 other hazardous chemicals are being vented into the atmosphere. One such is methyl isocyanate, which killed nearly 3000 people in Bhopal, India, from an industrial accident in 1984. Although there is yet no conclusive evidence concerning the long-term exposure of low levels of toxic substances in the atmosphere, there are some alarming signs, especially when these substances are rained out and end up in rivers, lakes, and soils.

RAINING ACID

One of the most studied and yet least understood problems of our modern industrial era is the spectacle of acid rain. In many parts of the world, architectural treasures are eroding and soil and lakes have become abnormally acidic, causing vegetation and fish populations to dwindle or disappear. In the most extreme cases, the rain has had the acidity of vinegar—a pH value of roughly 2 to 3. (The pH scale runs from 0, the strongest acid, to 14, the strongest base. A pH of 7 is neutral.)

The source of the problem has been identified for decades, leaving little doubt about the basic chemistry that turns industrial and automobile emissions into acid rain. The damage inflicted by acid precipitation led to the passage of the 1980 Acid Precipitation Act in the United States. This law prompted more studies. Since then, a great deal of information has been amassed. In some instances, the effects of acid rain have turned out to be less serious than it was feared, but in others, environmental factors seem to have made the problem much worse.

Scientists have learned that there are many components acting simultaneously to destroy the forests and lakes, and not just acid rain alone. The action of acid rain in concert with other pollutants such as metals, surface ozone, and various organic compounds places far greater stress on natural systems than was ever thought before.

Much of the environmental damage is caused by the direct precipitation of acids and dry deposits of sulfates. Coal-fired plants are the biggest producers of these acids and sulphates, and coal consumption has increased 70 percent since 1975. Although sulfur dioxide emissions from these facilities have also dropped some 10 percent during the same period, they still by far outweigh natural sources of sulfur dioxide such as volcanic eruptions.

When the sulfur dioxide gas encounters water or related molecules in the atmosphere, it yields sulfuric acid by the formula: $SO_2 + H_2O = H_2SO_4$. Nitric acid is produced by the reaction of moisture in the atmosphere with nitrogen oxide gas from the high-temperature combustion of fossil fuels in coal-fired plants and internal combustion engines by the formula: $NO_2 + H_2O = HNO_3$.

These acids are readily dissolved in rainwater, which runs off into streams and lakes and percolates into the soil. Fine plant roots are severely damaged by soil acids, which also leach valuable nutrients from the soil. The direct contact of acids on foliage also destroys crops, as well as forests. Fish, which have a particularly low tolerance to high acidity levels, are dying in large numbers, leaving many lakes in the northeastern United States totally devoid of fish. Those in the West appear not to be as badly affected. In addition, 14,000 lakes in Canada are threatened and more than 15,000 lakes in Sweden are now without fish because of acid rain.

In addition to acid rain, there is also acid snow, acid fog, and even acid dew. Although acid dew does not rival acid rain as an environmental hazard, it can be harmful. Its acidification process might be more complex than was suspected, which might play a very critical role in harming trees. The acidic dewdrops are not usually harmful at night, but their evaporation after the Sun rises tends to concentrate the acids, which could cause damage to the leaf surface upon which the dew condenses.

Acid dew forms when dewdrops absorb atmospheric nitric acid gas and sulfur dioxide, which

Table 14-1. Summary of Soil Types.

CLIMATE	TEMPERATE (HUMID) > 160 IN. RAINFALL	TEMPERATE (DRY) < 160 IN. RAINFALL	TROPICAL (HEAVY RAINFALL)	ARCTIC OR DESERT
VEGETATION	Forest	Grass and brush	Grass and trees	Almost none; no humus development
TYPICAL AREA	Eastern U.S.	Western U.S.		
SOIL TYPE	Pedalfer	Pedocal	Laterite	
TOPSOIL	Sandy, light colored; acid	Enriched in calcite; white color	Enriched in iron and aluminum, brick red color	No real soil forms because no organic material; chemical weathering very low
SUBSOIL	Enriched in aluminum, iron, and clay; brown color	Enriched in calcite; white color	All other elements removed by leaching	
REMARKS	Extreme development in conifer forest; abundant humus makes groundwater acid; soil light gray due to lack of iron	Caliche (name applied to accumulation of calcite)	Apparently bacteria destroy humus; no acid available to remove iron	

then become oxidized into sulfuric acid. It is also caused by the dry deposition of acid particles and gases settling on wet surfaces.

Roughly one-third of the sulfur dioxide produced in the United States reaches the ground by way of dry deposition, and it is suspected that dry deposits are as destructive to the environment as acid rain or snow. Sulfates are known to contribute most of the fine-particle mass over much of the eastern United States and many other regions. The sulfate particles are often highly acidic and could possibly damage materials and alter the acid-base balance as much as acid rain does.

THE POLLUTION SOLUTION

The hydrologic cycle serves to cleanse the Earth of natural and man-made pollutants through the process of dilution. When the system becomes overloaded with toxic wastes, however, there is the danger that these toxins will be concentrated to lethal levels. Many toxic substances that are diluted

supposed safe levels in streams, lakes, and oceans become concentrated by biological activity, starting at the bottom of the food chain and working its way up to fish and other aquatic life, which are then eaten by humans. Mercury poisoning of fish by industrial wastes dumped into rivers in Japan and DDT poisoning of penguins in the Antarctic worked through just such a process.

Despite environmental pressures, industrial toxic chemicals are still dumped into rivers and streams, either deliberately or accidentally, by industrialized nations. The worst accidental spill in the past decade occurred on the Rhine in November 1986, when a Swiss chemical factory caught fire and hundreds of tons of fungicides, pesticides, and other agricultural chemicals were washed into the river, producing a 25-mile-long chemical slick that killed about one-half million fish and eels over a stretch of 100 miles or more. Similarly, agricultural chemicals draining off Denmark and Sweden have so polluted the Kattegat, the sea between Sweden and Denmark, that approximately 30 million fish have died, making this possibly the greatest environmental catastrophe of the century.

Untreated sewage is dumped directly into the ocean because sewage treatment facilities are overloaded and funds are unavailable for improvements. As a result, swimming is unsafe in beaches in many parts of the world. So much garbage is generated in the big cities that the only place left to put it is into the sea. Unfortunately, ocean currents bring the wastes back to shore, and wastes tend to concentrate between thermal layers and ocean fronts, which also happen to be the feeding grounds for fish. In the Atlantic, the meandering currents of the Gulf Stream, which are laden with fish upon which the fishing industry greatly depends, actually sweep over the dump sites.

What is even more insidious is the incineration on the open sea of highly toxic substances that are considered too dangerous to detoxify on land. It is still highly uncertain what effects this technology will have on the marine environment.

Oil spills are by far the most damaging of all coastal pollution, and the increased demand for offshore oil (FIG. 14-3), the collisions of oil tankers, and the attacks on oil tankers by warring countries have led to disastrous ecological consequences. Heavy spills often require extensive cleanup efforts, especially in productive fishing grounds.

What is worse, marine pollution is not localized in highly contaminated areas such as the Mediterranean and the North Sea. Instead, the contaminants are spread to other parts of the world through the action of ocean currents.

One of the most important sources of water is groundwater. In the western and midwestern states, most of the water comes from the ground for domestic, industrial, and agricultural use. Pollutants from landfills, industrial wastes, agricultural chemicals, and even low-level nuclear wastes are finding their way into aquifers and contaminating what were once nearly pure sources of water that normally did not require treatment. Some of these contaminants are

(Courtesy of U.S. Maritime Administration)

FIG. 14-3. Oil tanker taking on crude oil from Alaska's North Slope at the port of Veldez.

carcinogenic, and even low levels of these substances can cause cancer.

Many wells are now so polluted that they must be treated at the well head, which depending on the extent of contamination can be expensive. Presently, up to 10 percent of the groundwater supply in the United States is already contaminated, and in the coming years, a quarter or more of the groundwater might be unusable. The pollutants flow so slowly, they can take several years before they even show up in the wells.

Cleaning up this contamination is expensive, difficult, and in some cases impossible. It requires locating the pollutant, which is not an easy task, drilling a well, and pumping the polluted water out. Sometimes if the pollutants are fairly localized, they can be blocked by pumping impermeable substances like clay into the aquifer. Unfortunately for much of the groundwater, past mistakes might have already made recovery too late.

LOSING GROUND

We are losing the soil beneath our feet faster than nature is putting it back. In a world where three-quarters of the people go to bed hungry, we can ill afford to destroy the very soil needed to sustain the human race even at present population levels. Throughout the world, most of the arable land is already under the plow, and efforts to cultivate substandard soils is leading to poor productivity and ultimately abandonment, which in turn leads to severe soil erosion. The soil profile (FIG. 14-4) is divided into the four zones. The A zone is a thin zone from a few inches to a few feet thick with an average of only 7 inches worldwide. It is where most of the soil nutrients are. The B zone below it is of poor soil quality. The effect of soil erosion is to bring the B zone to the surface. Because plants do not grow well in the subsoil, plant roots cannot hold it in and it too is eroded.

It takes a long time for rocks to be broken down into soil, and presently, the topsoil is disappearing up to three times faster than new soil is being generated. The soil is either blown away or carried away by streams (FIG. 14-5) and ultimately ends up on the bottom of a reservoir (FIG. 14-6) or the ocean. As

the world becomes more crowded and poor land is continually being planted and eroded, human beings are rapidly becoming the greatest geological force on the face of the Earth.

Desertification is a process that results from the loss of fine topsoil, leaving only the coarse sand behind after the land is denuded by improper farming methods and overgrazing. The once lush Fertile Crescent, which once fed as many as 25 million people, is now a desert because of poor farming practices in the past. The once great forests of the Sahel region of central Africa were destroyed by nomads to improve hunting and to provide grassland for herds. Now, the Sahel is being overrun by the sands of the Sahara desert, which are advancing inexora-

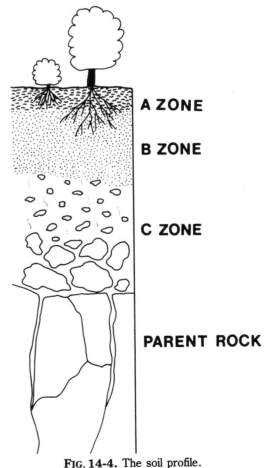

FIG. 14-4. The soil profile.

bly, engulfing everything in their path and chasing people out of the region.

The desertification process in exacerbated by the fact that without vegetation, the land is subjected to flash floods, higher rates of evaporation, and dust storms (FIG. 14-7). The denuded land also has a higher surface reflectance of sunlight, which contributes to lesser rainfall and in the long run, denudes more land.

Thus, man-made deserts march across what were once fertile lands. Throughout the world, perhaps as much as one-third to one-half of what was once arable land is now rendered useless by erosion and desertification. In addition, one-half to three-quarters of all irrigated land will be destroyed by salt buildup in the soil by the end of this century. Unless these trends are reversed, the future of mankind and much of the rest of the world remains highly uncertain.

THE FELLING OF THE FORESTS

Every year, tropical rain forests are being destroyed at a rate of about 40,000 square miles, or about the size of the state of Ohio. If the present global deforestation continues, the world's rain forests will be all but obliterated sometime during the next century. The forests are disappearing through small-scale slash-and-burn agriculture on one extreme and large-scale timber harvesting on the other.

Much of the soil underlying rain forests is thin and of poor quality. The ashes from burned trees can help fertilize the soil, but after a couple of years of farming, the soil becomes depleted in nutrients, and farmers are forced to press farther into the forest.

Timber companies use modern timber-harvesting equipment with giant shears that can snip huge trees off at the base and wood chippers that can reduce a 100-foot tree in just 30 seconds. Some of the rain forests are cut down for electrical power generation, which is a poor use of a valuable resource. Unwanted trees and brush are then burned and the bare soil, now totally denuded of all vegetation, is left unattended as the timber companies continue to mow down the forests.

Rain forests are so named because they receive 200 inches and more of rain each year. When the rains come, the denuded soil is washed away by fierce flash floods, leaving mostly bare rock behind. Without the soil, the chances of recovery are next to none, and for all practical purposes, the forests are gone for good.

Forests are extensive (FIG. 14-8) and act as climate controllers. Desertification brought on by the loss of the forests increases surface albedo, and sunlight is reflected out to space. The loss of solar energy could change precipitation patterns with a consequential decrease of rainfall, particularly in the rain forests, which could cause stress and make trees more susceptible to disease.

Soot from forest fires absorbs solar radiation, which then goes to heat the atmosphere. The result is a temperature imbalance, in which the temperature rises with altitude instead of falling like it should. Large quantities of soot in the atmosphere could, therefore, send abnormal weather patterns to all parts of the world.

Forests also store a great deal of carbon, and the clearing of forested land for agriculture, especially in the tropics, is the largest source of carbon released into the atmosphere. Although there is a net accumulation of carbon in the forests of North America and Europe, it is insignificant compared with the losses in the tropical regions. The increased carbon dioxide content in the atmosphere heightens the greenhouse effect, which can substantially change global weather patterns as the world heats up.

Deforestation also poses a threat to the ozone layer by releasing nitrous oxide into the atmosphere. Clear-cutting of timber encourages soil bacteria to produce nitrous oxide, which is then emitted into the air. The burning of the trees themselves also produces significant amounts of nitrous oxide, which might be a substantial source of this gas in the global atmosphere.

By far, the most damaging aspect of deforestation is the loss of wildlife habitat. Tropical rain forests cover only about 6 percent of the land surface, but they contain 60 percent or more of the existing plant and animal species. These species are continuously being crowded out by the encroachment of humans

FIG. 14-5. Severe streambank erosion along Muddy Creek, Cascade County, Montana.

FIG. 14-6. Sediment entering Lake Tahoe from Trout Creek and Upper Trukee River.

The Felling of the Forests **189**

Fig. 14-7. Duststorm in New Mexico.

into their habitat, and the consequential destruction of ecological niches and pollution of their environment with herbicides, insecticides, and industrial poisons. Some exotic plants that are in danger of becoming extinct have important medicinal value, and it would be criminal to lose those species that could possibly help in the fight against diseases such as AIDS and cancer.

POPULATION AND HUNGER

Two million years ago, the entire human population totaled just 100,000. By the end of the last ice age about 10,000 years ago, there were upwards of 10 million people in the world. Just 5,000 years ago, that number rose to 100 million. Around 1800, the population reached the 1 billion mark. By 1930,

that figure doubled to 2 billion. In 1960, after only 30 more years, the world's population reached 3 billion. Fifteen years later, another billion people were added. By 1987, after another dozen years, the population grew to 5 billion. If present trends continue, the population will double again during the first quarter of the next century, with 80 percent of the increase occurring in developing countries.

The nineteenth century English economist Thomas Malthus observed that human populations are limited by their food supply because they grow geometrically, while food production increases only arithmetically. Therefore, unless some constraints are placed on the burgeoning human population, there will be increasing misery, especially among the poor nations, which because of the destruction of

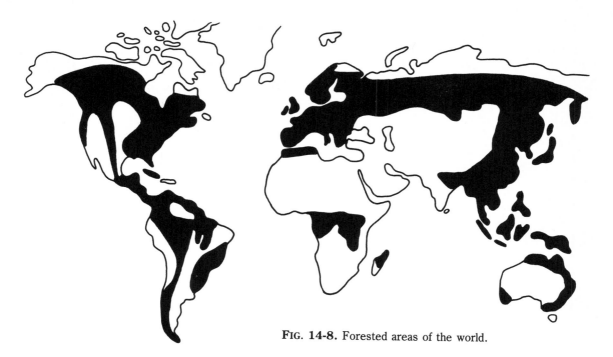

FIG. 14-8. Forested areas of the world.

their land have already surpassed the carrying capacity and can only survive with outside aid.

The leading exporters of food for the rest of the world already have most of their arable lands in production. During the decade of the seventies, American farmers placed an additional 60 million acres under cultivation—a larger area than the state of Kansas—in order to help feed a hungry world. Much of this increased land was substandard, including sloping, marginal, and fragile soils that were easily erodible. As a result, in 1977 alone, some 3 billion tons of soil were eroded.

Genetic engineering, thought to be a panacea to save the human race, has produced highly productive and drought-resistant varieties of plants. However, without large quantities of expensive fertilizers to restore the land and pesticides to ward off pests, third-world countries do not stand much of a chance, no matter what they plant. In addition, the improper use of agricultural chemicals can be exceedingly harmful to the environment. Two or more years of successive drought (FIG. 14-9) can wipe out any meager gains toward feeding the world's needy, with

the additional problem of greater desertification.

Further problems arise from political and religious ideologies, warfare, disease, and depletion of

FIG. 14-9. Drought-prone region of Africa.

FIG. 14-10. Sunset at the Bighorn Mountains, Wyoming.

natural resources. In many parts of the world, big families are desirable in order to help cultivate the land. The work is mostly done by manual labor, in which one calorie is expended in work for every food calorie. Often the food is of poor nutritional value, keeping many people just above starvation levels. Also, damage to the intestinal tract caused by parasites and infection makes the body less efficient in utilizing food, requiring a higher caloric intake in order to stay alive. Young children are particularly at risk, and those who survive mortality generally are diseased, stunted, blinded, or retarded. These effects are often irreversible. Therefore, not only is the population growing out of control, but the human condition is deteriorating at an alarming rate. In other words, increasing the quantities of human beings not only decreases the quality of life, but weakens stamina and the ability to ward off disease.

It is also becoming more apparent that humans are upsetting the delicate balance of nature, and if that balance should tilt ever so slightly one way or the other, cataclysmic changes could result. Over a period of more than 4.5 billion years, the Earth has perfected a means of survival for all living things. There have been major readjustments followed by major extinctions, and life generally came out of these tragic events better off than before.

Through the long process of evolution, each step up the ladder produced more advanced species, as though nature were guided by a will of its own. We know this is not so, and that life is simply a machine, albeit a highly complex one, that is compelled to follow natural laws. Therefore, on this planet, as well as other earthlike planets if they exist, the ultimate outcome of evolution should be an intelligent being that is capable of making logical choices concerning himself and his environment. He will recognize that he cannot survive if he puts his planet in jeopardy. Seeing that there is a serious problem, an intelligent being goes about in a rational way to correct it. The question remains whether human beings are truly intelligent enough to surmount the most perplexing problems ever encountered by the living Earth.

Bibliography

MAKING A SPECIAL PLANET

Bok, Bart J. "The Milky Way Galaxy." *Scientific American* Vol. 244 (March 1981): 92-120.

Boss, Alan P. "The Origin of the Moon." *Science* Vol. 231 (January 24, 1986): 341-345.

Herbst, William and George E. Assousa. "Supernovas and Star Formation." *Scientific American* Vol. 241 (August 1979): 138-144.

Kerr, Richard. "Continents at the Core-Mantle Boundary?" *Science* Vol. 233 (August 1, 1986): 523-524.

Maddox, John. "Origin of Solar System redefined." *Nature* Vol. 308 (March 15, 1984): 223.

Mulholland, Derral. "The Beast at the Center of the Galaxy." *Science 85* Vol. 6 (September 1985): 51-55.

O'Nions, R.K., P.J. Hamilton, and Norman M. Evensen. "The Chemical Evolution of the Earth's Mantle." *Scientific American* Vol. 242 (May 1980): 120-133.

Schramm, David N. and Robert N. Clayton. "Did a Supernova Trigger the Formation of the Solar System?" *Scientific American* Vol. 239 (October 1978): 124-139.

Thompson, Dietrick, E. "A very old galaxy may be very young." *Science News* Vol. 131 (January 10, 1987): 23.

Weisburd, Stefi. "The Inner Earth Is Coming Out." *Science News* Vol. 131 (April 4, 1987): 222-223.

Wetherill, George W. "The Formation of the Earth from Planetesimals." *Scientific American* Vol. 224 (June 1981): 163-174.

THE CARBON CONNECTION

Cairns-Smith, A. G. "The First Organisms." *Scientific American* Vol. 252 (June 1985): 90-100.

Cloud, Preston. "The Biosphere." *Scientific American* Vol. 249 (September 1983): 176-189.

Gilmore, V. Elaine. "Did life begin in clay?" *Popular Science* Vol. 227 (September 1985): 32.

Holland, H. D., B. Lazar, and M. McCaffrey. "Evolution of the atmosphere and oceans." *Nature* Vol. 320 (March 6, 1986): 27-33.

Ingersoll, Andrew P. "The Atmosphere." *Scientific American* Vol. 249 (September 1983): 162-174.

Lewin, Roger. "RNA Catalysis Gives Fresh Perspective on the Origin of Life." *Science* Vol. 321 (February 7, 1986): 545-546.

Raloff, Janet. "Is there a cosmic chemistry of life?" *Science News* Vol. 130 (September 20, 1986): 182.

Stokes, W. Lee. *Essentials of Earth History*. Prentice-Hall, 1982.

Toon, Owen B. and Steve Olson. "The Warm Earth." *Science 85* Vol. 6 (October 1985): 50–57.

Weinberg, Robert A. "The Molecules of Life." *Scientific American* Vol. 253 (October 1985): 48–57.

AGE OF EARLY LIFE

Attenborough, David. *Life on Earth*. Little, Brown, 1979.

Brock, Thomas D. "Precambrian Evolution." *Nature* Vol. 288 (November 20, 1987): 214–215.

Dickerson, Richard E. "Cytochrome c and the Evolution of Energy Metabolism." *Scientific American* Vol. 242 (March 1980): 137–153.

Fisher, Arthur. "Bacterial missing link." *Popular Science* Vol. 230 (June 1987): 12.

Ford, Trevor D. "Life in the Precambrian." *Nature* Vol. 285 (May 22, 1980): 193–194.

Groves, David I., John S. R. Dunlop, and Roger Buick. "An Early Habitat of Life." *Scientific American* Vol. 245 (October 1981): 64–73.

Kerr, Richard A. "Plate Tectonics Is the Key to the Distant Past." *Science* Vol. 234 (November 7, 1986): 670–672.

McMenamin, Mark A. S. "The Emergence of Animals." *Scientific American* Vol. 256 (April 1987): 94–102.

Valentine, James W., and Eldridge M. Morres. "Plate Tectonics and History of Life in the Oceans." *Scientific American* Vol. 230 (April 1974): 80–89.

Youvan, Douglas C., and Barry L. Marrs. "Molecular Mechanics of Photosynthesis." *Scientific American* Vol. 256 (June 1987): 42–48.

AGE OF ANCIENT LIFE

Eldredge, Niles. *Life Pulse; Episodes from the Story of the Fossil Record*. Facts on File, 1987.

Gosline, John M. and M. Edwin DeMont. "Jet-propelled Swimming in Squids." *Scientific American* Vol. 252 (January 1985): 96–103.

Lewin, Roger. "Computers Track the Path of Plant Evolution." *Science* Vol. 219 (March 11, 1983): 1203–1205.

———. "On the Origin of Insect Wings." *Science* Vol. 230 (October 25, 1985): 428–429.

Mintz, Leigh W. *Historical Geology; The Science of a Dynamic Earth*. Merrill, 1972.

Morris, Simon Conway, and H. B. Whittington. "The Animals of the Burgus Shale." *Scientific American* Vol. 241 (July 1979): 122–133.

Niklas, Karl J. "Computer-simulated Plant Evolution." *Scientific American* Vol. 254 (March 1986): 78–86.

AGE OF MIDDLE LIFE

Bakker, Robert T. "Dinosaur Renaissance." *Scientific American* Vol. 232 (April 1975): 58–78.

Buffetaut, Eric. "The Evolution of the Crocodilians." *Scientific American* Vol. 241 (October 1979): 130–144.

Emery, Kenneth. "Precursors of Mammals." *Science Digest* Vol. 90 (January 1982): 19.

Horner, John R. "The Nesting Behavior of Dinosaurs." *Scientific American* Vol. 250 (April 1984): 130–137.

Kerr, Richard A. "How to Make a Warm Cretaceous Climate." *Science* Vol. 223 (February 17, 1984): 677–678.

Langstrom, Wann, Jr. "Pterosaurs." *Scientific American* Vol. 244 (February 1981): 122–136.

Leg 89 Staff. "The Mesozoic superocean." *Nature* Vol. 302 (March 31, 1983): 381.

Morell, Virginia. "Announcing the Birth of a Heresy." *Discover* Vol. 8 (March 1987): 26–50.

Mossman, David J., and William A. S. Sarjeant. "The Footprints of Extinct Animals." *Scientific American* Vol. 248 (January 1983): 75–85.

AGE OF RECENT LIFE

Begley, Sharon. "The Mystery of the Whales." *Newsweek* (October 22, 1984): 52.

Benton, J. Michael. "First marsupial fossil from Asia." *Nature* Vol. 318 (November 28, 1985): 313.

Bower, Bruce. "Fossil Finds Diversify Ancient Apes." *Science News* Vol. 130 (November 22, 1986): 324.

———. "An ancient relative for the owl monkey." *Science News* Vol. 131 (April 25, 1987): 263.

Gould, James L. and Peter Marler. "Learning by Instinct." *Scientific American* Vol. 256 (January 1987): 74–85.

Jerison, Harry J. "Paleoneurology and the Evolution of the Mind." *Scientific American* Vol. 234 (January 1976): 90–101.

Kanwisher, John W. and Sam H. Ridgway. "The Physiological Ecology of Whales and Porpoises." *Scientific American* Vol. 248 (June 1983): 111–120.

Lewin, Roger. "Why Is Ape Tool Use So Confusing?" *Science* Vol. 236 (May 15, 1987): 776–777.

Weisburd, Stefi. "Oldest marsupial fossil Found?" *Science News* Vol. 129 (May 10, 1986): 295.

THE BIG BRAIN

Begley, Sharon and Louise Lief. "The Way We Were." *Newsweek* (November 10, 1986): 62–72.

Bower, Bruce. "Skull Gives Hominid Evolution New Face." *Science News* Vol. 130 (August 16, 1986): 100.

_____. "When The Human Spirit Soared." *Science News* Vol. 130 (December 13, 1986): 378–379.

Brown, Frank, John Harris, Richard Leakey, and Alan Walker. "Early Homo erectus skeleton from west Lake Turkana, Kenya." *Nature* Vol. 316 (August 29, 1985): 788–792.

Burian, Zdenek. *The Dawn of Man*. Thames and Hudson, 1978.

Diamond, David. "The Worst Mistake in the History of the Human Race." *Discover* Vol. 8 (May 1987): 64–66.

Lewin, Roger. "Four Legs Bad, Two Legs Good." *Science* Vol. 235 (February 27, 1987): 969–971.

_____. "Man the Scavenger." *Science* Vol. 224 (May 25, 1984): 861–862.

Pilbeam, David. "The Descent of Hominoids and Hominids." *Scientific American* Vol. 250 (March 1984): 84–96.

Rukang Wu and Lin Shenglong. "Peking Man." *Scientific American* Vol. 248 (January 1983): 86–94.

Trinkaus, Erik and William W. Howells. "The Neanderthals." *Scientific American* Vol. 241 (December 1979): 118–133.

THE LIFE CYCLES

Covy, Curt. "The Earth's Orbit and the Ice Ages." *Scientific American* Vol. 250 (February 1984): 58–66.

Gough, Douglas. "What Causes the solar cycle?" *Nature* Vol. 319 (January 23, 1986): 263–264.

Kerr, Richard A. "Monitoring Earth and Sun by Satellite." *Science* Vol. 236 (June 26, 1987): 1624–1625.

_____. "The Moon Influences Western U.S. Drought." *Science* Vol. 224 (May 11, 1984): 587.

Meko, D.M., C.W. Stockman, and T.J. Blasing. "Periodicity in Tree Rings from the Corn Belt." *Science* Vol. 229 (July 26, 1985): 381–384.

Palmer, John D. "Biological Clocks of the Tidal Zone." *Scientific American* Vol. 232 (February 1975): 70–79.

Pittock, A. Barrie. "Cycles in the Precambrian." *Nature* Vol. 318 (December 12, 1985): 509–510.

Siever, Raymond. "The Steady State of the Earth's Crust, Atmosphere, and Oceans." *Scientific American* Vol. 230 (January 1974): 72–79.

Thomsen, Dietrick E. "A more complex solar cycle." *Science News* Vol. 131 (January 17, 1987): 39.

Willians, George E. "The Solar Cycle in Precambrian Time." *Scientific American* Vol. 255 (August 86): 88–96.

ICE ON THE WORLD

Andrews, J. T. "Short ice age 230,000 years ago?" *Nature* Vol. 303 (May 5, 1983): 21–22.

Bowen, D. Q. "Antarctic ice surges and theories of glaciation." *Nature* Vol. 283 (February 14, 1980): 619–621.

Campbell, Philip. "New data upset ice age theories." *Nature* Vol. 307 (February 23, 1984): 688–689.

Fodor, R. V. "Explaining the Ice Ages." *Weatherwise* (June 1982): 109–114.

Kerr, Richard A. "An Early Glacial Two-Step." *Science* Vol. 221 (July 8, 1983): 143–144.

_____. "Climate Since the Ice Began to Melt." *Science* Vol. 226 (October 19, 1984): 326–327.

Matthews, Samuel W. "Ice On the World." *National Geographic* (January 1987): 79–103.

Moore, Peter D. "Clues to past climate in river sediment." *Nature* Vol. 308 (March 22, 1984): 316.

Paresce, Francesco and Stuart Bowyer. "The Sun and the Interstellar Medium." *Scientific American* Vol. 255 (September 1986): 93–99.

Weisburd, Stefi. "Forests made the world frigid?" *Science News* Vol. 131 (January 3, 1987): 9.

THE GREAT HEAT ENGINE

Ambroggi, Robert P. "Water." *Scientific American* Vol. 243 (September 1980): 101–115.

Barber, Richard T. and Francisco P. Chaves. "Biological Consequences of El Niño." *Science* Vol. 222 (December 16, 1983): 1203–1210.

Borecker, Wallace S. "The Ocean." *Scientific American* Vol. 249 (September 1983): 146–160.

Friend, P.F. "Storms in the Abyss." *Nature* Vol. 309 (May 17, 1984): 212.

Hollister, Charles D., Arthur R. M. Nowell, and Peter A. Jumars. "The Dynamic Abyss." *Scientific American* Vol. 250 (March 1984): 42–53.

Kerr, Richard A. "Are the Ocean's Deserts Blooming." *Science* Vol. 220 (April 22, 1983): 397–398.

_____. "Small Eddies Are Mixing the Oceans." *Science* Vol. 230 (November 15, 1985): 793.

MacIntyre, Ferren. "The Top Millimeter of the Ocean." *Scientific American* Vol. 230 (May 1974): 62–77.

Neilson, Ronald P. "High-Resolution Climatic Analysis and Southwest Biogeography." *Science* Vol. 232 (April 4, 1986): 27–33.

Ramage, Colin S. "El Niño." *Scientific American* Vol. 254 (June 1986): 77–83.

Silberner, Joanne. "U.S. weather waxing cloudy." *Science News* Vol. 131 (March 28, 1987): 200.

Wabster, Peter J. "Monsoons." *Scientific American* Vol. 245 (August 1981): 109–118.

THE SUPERCONTINENT

Bonatti, Enrico. "The Rifting of Continents." *Scientific American* Vol. 256 (March 1987): 97–103.

Courtillot, Vincent and Gregory E. Vink. "How Continents Break Up." *Scientific American* Vol.

249 (July 1983): 43–49.

Fisher, Arthur. "Alaska down under?" *Popular Science* Vol. 228 (June 1986): 10–11.

Hekinian, Roger. "Undersea Volcanoes." *Scientific American* Vol. 251 (July 1984): 46–55.

Kerr, Richard A. "Continental Drift Nearing Certain Detection." *Science* Vol. 229 (September 6, 1985): 953–955.

_____. "Do Tectonic Plates Drive Themselves?" *Science* Vol. 236 (June 12, 1987): 1426–1427.

Mutter, John C. "Seismic Images of Plate Boundaries." *Scientific American* Vol. 254 (February 86): 66–75.

Sclater, John G. and Christopher Tapscott. "The History of the Atlantic." *Scientific American* Vol. 240 (June 1979): 156–174.

Weisburg, Stefi. "India gets under Eurasia's skin." *Science News* Vol. 130 (October 18, 1986): 253.

LIFE IN STRANGE PLACES

Brock, Thomas D. "Life at High Temperatures." *Science* Vol. 230 (October 11, 1985): 132–138.

Childress, James J., Horst Felbeck, and George N. Somero. "Symbiosis in the Deep Sea." *Scientific American* Vol. 256 (May 1987): 115–120.

Conover, R. J., A. W. Herman, S. J. Prinsenberg, and L. R. Harris. "Distribution of and Feeding by the Copepod Pseudocalanus Under Fast Ice During the Arctic Spring." *Science* Vol. 232 (June 6, 1986): 1245–1247.

Eastman, Joseph T. and Arthur L. DeVries. "Antarctic Fishes." *Scientific American* Vol. 255 (November 1986): 106–114.

Edmond, John M. and Karen Von Damm. "Hot Springs on the Ocean Floor." *Scientific American* Vol. 248 (April 1983): 78–93.

Isaacs, John D. and Richard A. Schwartzlosc. "Active Animals of the Deep-Sea." *Scientific American* Vol. 233 (October 1975): 88–91.

Kerr, Richard A. "Ocean Hot Springs Similar Around Globe." *Science* Vol. 235 (January 23, 1987): 435.

Littler, Mark M. and Diane S. Littler. "Deepest Known Plant Life Discovered on an Uncharted Seamount." *Science* Vol. 227 (January 4, 1985): 57–59.

Penkett, S. A. "Implications of Arctic air pollution." *Nature* Vol. 311 (September 27, 1984): 299.

Steinhorn, Ilana, Joel R. Gat. "The Dead Sea." *Scientific American* Vol. 249 (October 1983): 102–109.

Weisburd, Stefi. "Halos of Stone." *Science News* Vol. 127 (January 19, 1985): 42–44.

EXTINCTIONS ARE A WAY OF LIFE

Benton, Michael J. "Interpretations of mass extinction." *Nature* Vol. 314 (April 11, 1986): 496–497.

Bower, Bruce. "Extinctions on Ice." *Science News* Vol. 132 (October 31, 1987): 284–285.

Jablonski, David. "Background and Mass Extinction: The Alternation of Macroevolutionary Regimes." *Science* Vol. 231 (January 10, 1986): 129–132.

Lewin, Roger. "Extinctions and the History of Life." *Science* Vol. 221 (September 2, 1983): 935–937.

———. "Mass Extinctions Select Different Species." *Science* Vol. 231 (January 17, 1986): 219–220.

———. "A Mass Extinction Without Asteroids." *Science* Vol. 234 (October 3, 1986): 14–15.

Maddox, John. "Extinction by catastrophe?" *Nature* Vol. 308 (April 19, 1984): 685.

Raup, David M. "Biological Extinction in Earth History." *Science* Vol. 231 (March 28, 1986): 1528–1533.

Raup, David M. and J. John Sepkoski, Jr. "Periodic Extinction of Families and Genera." *Science* Vol. 231 (February 21, 1986): 833–836.

Russell, Dale A. "The Mass Extinctions of the Late Mesozoic." *Scientific American* Vol. 256 (January 1982): 58–65.

Stanley, Steven M. "Mass Extinctions in the Ocean." *Scientific American* Vol. 250 (June 1984): 64–72.

Ward, Peter. "The Extinction of the Ammonites." *Scientific American* Vol. 249 (October 1983): 136–147.

LIFE HANGING IN THE BALANCE

Bascom, Willard. "The Disposal of Waste in the Ocean." *Scientific American* Vol. 231 (August 1974): 16–25.

Brown, Lester R. "World Population Growth, Soil Erosion, and Food Security." *Science* Vol. 214 (November 27, 1981): 995–1001.

Gibbons, Boyd. "Do We Treat Our Soil Like Dirt?" *National Geographic* Vol. 166 (September 1984): 353–388.

Holden, Constance. "A Revisionist Look at Population and Growth." *Science* Vol. 231 (March 28, 1986): 1493–1494.

Idso, S. B. "Industrial age leading to the greening of the Earth?" *Nature* Vol. 320 (March 6, 1986): 22.

Kerr, Richard A. "Halocarbons Linked to Ozone Hole." *Science* Vol. 236 (June 5, 1987): 1182–1183.

Lewin, Roger. "Damage to Tropical Forests, or Why Were There So Many Kinds of Animals." *Science* Vol. 234 (October 10, 1986): 149–150.

Likens, Gene E., Richard F. Wright, James N. Galloway, and Thomas J. Butler. "Acid Rain." *Scientific American* Vol. 241 (October 1979): 43–51.

Maranto, Gina. "The Creeping Poison Underground." *Discover* Vol. 6 (February 1985): 75–78.

Monastersky, Richard. "Acid dew: What it does." *Science News* Vol. 132 (October 17, 1987): 247.

Peterson, Ivars. "Watch on Acid Rain: A Midterm Report." *Science News* Vol. 132 (January 18, 1987): 36.

Raloff, Janet. "Deforestation: Major threat to ozone." *Science News* Vol. 130 (August 23, 1986): 119.

———. "Salt of the Earth." *Science News* Vol. 126 (November 10, 1984): 290–301.

Revelle, Roger. "Carbon Dioxide and World Climate." *Scientific American* Vol. 247 (August 1982): 35–43.

———. "Food and Population." *Scientific American* Vol. 231 (September 1974): 161–170.

Shaw, Robert W. "Air Pollution by Particles." *Scientific American* Vol. 257 (August 1987): 96–103.

Sun, Marjorie. "Ground Water Ills: Many Diagnoses, Few Remedies." *Science* Vol. 232 (June 20, 1986): 1490–1493.

Index